D0772284

The
Nature of
Spring

Jim
Crumley

Published by Saraband
Digital World Centre, 1 Lowry Plaza,
The Quays, Salford, M50 3UB
and
Suite 202, 98 Woodlands Road,
Glasgow, G3 6HB, Scotland
www.saraband.net

Editor: Craig Hillsley

ISBN: 9781912235377
ebook: 9781912235384

Printed and bound in Great Britain by Clays Ltd, Elcograf S.p.A.

10 9 8 7 6 5 4 3 2 1

Contents

Part One

Harbingers

Harbingers

Whiles, the jackies hud their wheesht,
an eerie cortege abune the birks, conversin wi' nocht
but the saft creak o' fower-hunner weengs,

the land's loudest whisper.

In its unchancy wake the hale Highland Edge
shiddered and stilled
wi' a whoosh o' silence.

That was winter breathed its last.

Then, at a signal frae the yorlin on the wire,
the first moon o' spring heaved up i' the east,
a thissell-cok skirled i' the gloamin,

while deep i' the Mediterranean sooth

far, far past France,
a gowk atop an olive tree
checked in wi' its biological clock

and thocht: "Aye, it maun be time".

whiles – sometimes	*yorlin* – yellowhammer
jackies – jackdaws	*thissell-cok* – mistle thrush
hud their wheesht – fall silent	*gowk* – cuckoo
birks – birches	*maun* – must
fower-hunner – four hundred	

Chapter One

First Syllables

I like to go outside and paint pictures in the early spring. I suppose it's my way of trying to be a tulip, pushing my way out of the tight white bulb of winter and opening a little color against the drabness.

Ted Kooser,
Seasons in the Bohemian Alps
(University of Nebraska Press, 2002)

THE LAND IS PATCHED with frozen and slowly thawing snow, depending on whether it lies in shadow or sunlight. Ben Ledi, this Stirlingshire landscape's symbolic pyramid, seen through a waving screen of bare birches, has been whitened and softened and curved and blunted by masses of snow that fleeced the mountains in the night. Far below, Loch Venachar lies in a *contre-jour* dwam, its surface mimicking sky colours – iciest blue, various greys, white and gold – and contriving to hurl fragments of all of these up the hillside, so that it looks as if sky and loch and land have been daubed by the same brush, that this portion of the Earth has been unified by nature in colour and texture and purpose. Paul Cézanne would do just that in his later years, flooding his canvasses of Mont Sainte-Victoire with daring bravado, and from foreground to middle ground to mountaintop and on into the sky, the whole patched with the same few shades of greens, yellows and blues, and leaving just enough canvas unpainted to insinuate flecks of brightness on mountaintop and cloud. He wrote then:

I become more lucid in front of nature…but I cannot attain the intensity which unfolds to my senses. I don't have the magnificent richness of colouration which animates nature.

And:

Art is a harmony parallel to nature.

And his biographer, Alex Danchev (*Cézanne – A Life*, Profile Books, 2012), wrote:

Harmony, like beauty, was being redefined.

So, on a February afternoon of can't-possibly-be-spring-yet, the ghost of Cézanne bestrides this living hillside between Loch Venachar and Ben Ledi, animating nature, redefining its harmony and beauty by flooding its canvas with a limited palette that does not discriminate from foreground to loch to mountaintop and on into the sky. The result is a landscape fizzing with energy. But is it spring?

A bare, half-folded-fan-shaped rowan tree cuts through a low hill skyline so that its many-branched crown arcs against the sky and reveals the neat and placid silhouette of a vertically perched kestrel clasped to the topmost inch of the topmost twig. For the next five minutes it is utterly motionless (and for how long before I first saw it?), while the sky behind and above jigs and reels where wind and sun and cloud and blue and unpainted scraps of canvas conspire to create the appropriate setting to show it off. But the kestrel's preoccupation is the land below the tree, and that land is all pale gold and straw-shaded grasses, liberally patched with snow that gleams with hints of all those sky shades.

The bird throws its wings wide, raises its tail, and a shudder goes through it, and when that subsides it has realigned its stance from vertical to horizontal, and in the process it has revealed itself as a female kestrel, and now she too has redefined harmony and beauty. She will fly any moment.

She flies west. West is my direction, too, but more slowly now, for she is ahead of me, and I would like to encounter her again as we trace our parallel contours across the hillside.

She is not hard to find. She's there beyond the next small rise in the land. She perches on air now. Every feather and nerve-end is a-flicker, apart from her head, which is the still centre of her world. She steps down a yard from her perch, she banks, chooses a new course, urges forward into the west wind. I never saw a raptor that did not prefer to hunt into the wind. It is as true of thrush-sized merlins as it is of eagles, as it is of kestrels. Facing the wind achieves control. A tail wind brings chaos, unless the bird's ambition is to cover a lot of ground without the need for control. A hunting kestrel is precise. She will wheel onto a tail wind only when she reaches her own idea of the edge of her territory or her comfort zone, and then she may well surge all the way back to where she started, and there, if she's still hungry, she may turn and start again.

She keeps pace with me for about a mile. Several times I will overtake her because she is head-down in mid-hover, but sooner or later she reappears going west, going past me, always below me, always rising somewhere beyond to hold up against the sky again, intimately coupled to the wind. Finally, fruitlessly, she whips round in not much more than her own length and barges away east, downwind.

The path is comparatively new and has made a neat, deliberate breach in a drystane dyke that crosses it at right angles. It is not the first breach, and they are not all this neat, for the dyke wears the forlorn expression of the redundant and disregarded. But it inclines away uphill towards a wooded little crag down which shivers a muttering burn, and it reads like an invitation: "This way." The snow patches lie thicker there, ensnared by tussocks, a neatly stitched seam all the way up the shadowed side of the dyke. The burn's voice grows louder, blends into the fractured, percussive rhythms of a half-hearted waterfall hidden somewhere up there in the trees that crowd the crag and the low ground beneath. This hillside is full of these little dens, the gorges of their burns steep enough to discourage the grazing tribes throughout the centuries of careless farming that held sway hereabouts. But this land is in the hands of Woodland Trust Scotland now, and evidence of new native woodland planting is everywhere. A long, slow process of healing has begun, which is why I come often. So much of our relationship with nature is conducted in an atmosphere of battlefield; I greet these reprieved and resurrecting acres with over-emotional regard. The forgiveness of nature always moves me. It rises to accept this extended helping hand, and its response swarms all over this land, a freewheeling generosity.

Up near the crag, the dyke is taller and more intact. A level square yard of hillside at its base offers a seat, a backrest and a windbreak. My backpack offers coffee and biscuits. The sound of the unseen waterfall rushes into the void left by silenced footsteps and stillness. Settle for a while. The view faces east. A tall, slender birch of singular elegance distinguishes the middle distance. It leans away from its hillside, but then ten feet above the ground it begins to curve gently back until it achieves something slightly past vertical and (to my un-birch-like eye) perfect balance. My eye lingers there not just because of the beauty of the solitary tree, but also because the sun alights fitfully on a distant hill called Uam Mhor, and quite by chance my choice of a seat has put the birch and the hill on precisely the same sightline. But the sun is on the hill, and not the birch, so the vivid snow gleam of the hill appears from behind and therefore apparently from *within* the birch. The tree is lit from within.

What was it you said, M. Cézanne?

I become more lucid in front of nature…but I cannot attain the intensity which unfolds before my senses.

I think perhaps I know how you feel.

Then, quite unannounced, the unambiguous voice of spring: it creeps into the edges of consciousness as if from far off, but it cannot be far off. It infiltrates the predominance of the waterfall, even though it is much quieter, yet because its pitch is much higher and occasionally strident, it finds ways through the sound-screen of the fall, through the gaps in its acoustic limitations. Somewhere above the fall, high in the crag's highest trees, a mistle thrush sings. Has it just begun, or has it been singing all along and it has taken time for me to tune in to the changed circumstances created by the angle of the dyke and the crag? The song reaches me in short, staccato phrases, often tapering away to silence, one diminuendo after another, and further fragmented because from time to time the fall drowns it out. But listen. If you like your harbingers well-toned, jazzily inventive and far-carrying, accompanied by coffee *al fresco* and a birch tree performing passable impressions of the burning bush, and if you are willing to turn a blind eye or two to the snow-patched land...then here on this Cézanne-animated February late afternoon, these are the first syllables of spring.

The singer is an un-mated male. And while it's true that you can sometimes hear him on a fine afternoon of late December, and at any time in January, these are moments of overture. But this, this is song for the sake of song, an advertisement, yes, but also an outpouring of intent, a declaration that winter is lost and irretrievable now. It is the first day of mistle thrush spring.

My mood suffuses. It permeates hillside and drystane dyke (which moulders day by day at nature's prodding and urging, regressing back into hillside). It permeates tree, crag, waterfall, thrush, song. The process of grafting on to nature's late afternoon hour slowly creeps beyond mind, beyond senses, becomes physical, becomes bodily. The thing is to be *of the land*, to breathe in unison, to keep its peace.

On and on the thrush.

On and on the waterfall.

On and on the solitary birch, silvering the air.

These are the golden moments. They let me in, and briefly, I belong.

⊙ ⊙ ⊙

Something about the birch.

What?

What changed?

Why does it suddenly snap into sharp focus as if it has just wandered up the hill this moment, to stand there and adopt that leaning, curving pose? It has not moved, of course. It has not become more silvery. Its shadow has not deepened. But where its many-branched crown arcs against the sky, it reveals the neatly folded silhouette (clasped to the topmost inch of the topmost twig) of the very same vertically perched kestrel. And I never saw her arrive. How long has she been there?

On and on the fall.

On and on the thrush.

But the kestrel's reappearance unsettles me. It concludes a circle. And the knowledge creeps in with the thickening dusk that I belong beyond it.

Chapter Two

Falcons of the Yellow Hill

There is a curious relationship between peregrines and kestrels
that is difficult to define. The two species are often seen in the
same place, especially in autumn and spring. I rarely saw one
of them without finding the other close by. They may share the
same bathing places, the peregrine may occasionally rob the
kestrel of its prey, the kestrel may feed on kills the peregrine
has left, the peregrine may attack birds the kestrel unwittingly
puts up for him. In September and October some peregrines
seem to copy the kestrel's way of hunting, and I have seen the
two species hover together over the same field. In a similar
way, I have seen a peregrine hunting near a short-eared owl,
and apparently mimicking its style of flight. By March the
relationship between kestrel and peregrine has changed; the
peregrine has become hostile, and will stoop at, and probably
kill, any kestrel hovering near him.

J.A. Baker,
The Peregrine
(Collins, 1967)

THERE IS AN OLD Scottish ballad in which the young man
professes his love for the young woman thus: he will never

stop loving her until the whins stop flowering. You like to think the woman was botanically savvy enough to know that whins flower to one extent or another every month of the year. Otherwise, the singer is out of luck and there is no ballad. I mention it because on the last day of February, and from half a mile away, the whins still looked furled. They sprawled, impenetrable and dark green as stoorie old beer bottles, all the way from the roots of the big trees that fringed the bottom of the hill to the big crag that fringed the top. That sprawl was only stymied where buttress and scree intruded, where even whins fail to draw sustenance. But that view of the whins on their hillside from half a mile away was, like so much of these winter-into-spring days, something of a deception. Shrink the distance to a handful of yards and you would have found whin flowers star-bursting open in vivid yellow clusters all over the lower slopes, and that only the sheer dark mass of the still-unflowering bushes had masked the reality at a distance.

It was cold that day, the last day of meteorological winter; it snowed lightly, dustily, prettily, and every few minutes. And the coldness of the air brought to mind Margiad Evans, who wrote one of my favourite winter-wind images in all literature: "the wind is a tooth in the breast." Besides, there had been some very sinister weather forecasts, couched in a language (the "Beast from the East" – the Met Office had started talking like the *Daily Mail*, which was vaguely disconcerting) designed to attract the attention of the media so that no one should be in any doubt about what was coming our way from Siberia. Sadly, it worked, and weather-hysteria broke out. Cue panic-buying in the supermarkets. But for once, the reality more or less lived up to the hype. An impressive snow-storm was about to set in and last for several days, and unless you were a nature writer writing a book about the nature of spring, life would pretty well come to a standstill.

For the moment though, and before that freakish storm broke, I made a little hay while the sun shone between charmingly feathery snow showers, safe in the knowledge that whatever the nature of the Beast that lay in wait, come April the hillside would be a yellow blaze you could see for miles, as it was every April. Walking there, I would be able to get agreeably drunk on the smell of coconut, which is the whins' gift to the world. At least that was the theory, but more than thirty years of writing about nature in Scotland has taught me that it is a restless creature, that it thrives on variety, and that what I once thought of as reasonably set-tled patterns are being dismantled by an increasing tendency towards random and restless weather, itself a small microcos-mic symptom of an infinitely greater global climate chaos. Even keeping half an eye open for the vagaries of climate change teaches you that that very restlessness is on a quick-ening downhill trajectory at the end of which lies mayhem. The "Beast from the East" made crazy headlines while it lasted (it dumped eighteen inches of snow in my garden in thirty-six hours, which was exactly eighteen inches more than in the whole of the previous winter), but it was over and done with in a week, and in one week more the wind had returned to the west and temperatures were in double figures. But then it got cold again when spring should have sprung, and in my workaday landscape, everything stopped again, and that troubled this watcher of nature.

◉ ◉ ◉

From the foot of that whin-clad hillside, the skyline is a crag. It is half a mile long, broken only by a gully that splits almost the entire hillside, and that same entire hillside brims with birds in spring. These can drink and bathe in safety and with discretion in the gully's steep-tumbling burn, for it burrows deep once it escapes the gully, and whins and the hardiest of trees crowd round its lower reaches. The birds include not just the crag dwellers (peregrine falcon, kestrel, jackdaw, raven), but also the whin dwellers (robin, wren, dunnock, finch, warbler, chat, yellowhammer), and the woodland dwellers from the band of tall conifers and huge beeches that shields the foot of the hill from the road (sparrowhawk, owl, woodpecker, wood pigeon, blackbird, thrush, magpie, jay: it's a vociferous wood). Nature is profligate with its birds here, which in turn ensures the presence of a diversity of raptors – nesting kestrels, buzzards, ravens, those woodland sparrowhawks, and occasional red kites whose nest I have yet to pin down. And I would love to know how many foxes thread the whins and the skinny paths up through the rocky heights. They show themselves in sunlit pauses between whinbushes or tiptoeing across screes, glimpses of fiery red to complement the pervasive yellow. But the roost-rulers, high in the penthouse of the crag, have long been the per-egrine falcons, or so it has always seemed to me, but this perverse spring was about to challenge one or two of my most painstakingly composed theories.

This hillside is two miles away from my writing table, so I wander that way often and never return empty-handed.

And I have wondered from time to time if the peregrine that occasionally speculates above the gardens and woodland on the far side of the street is one of that crag's pair. My inclination thinks that it is. My reasoning is that if it were from a different and nearer nest, its raucous nesting-season voice would have betrayed the place by now. Besides, the crag-nesters could fly from there to here in less than five minutes, and I think that sometimes that's exactly what happens.

One of the fascinations of this particular pair of peregrines is that they have those nesting kestrels for near neighbours. They are separated by no more than 300 yards, which seems to be very near-neighbourly for two such virtuosic raptor-fliers. They can both hover, and they can both sprint, but the kestrel does hovering masterclasses that are beyond the skills of any peregrines I ever saw, while the power-diving peregrine is the fastest creature in this sky or any other, consigning the sprinting kestrel to a very distant second place. When they hunt, mostly they rely on their own specialisms.

Even more intriguing is that their territories overlap, so you might think that their explorations of the hunting possibilities of the same hillside and beyond are bound to get on each other's nerves a bit. With that in mind, I invite you to reconsider that passage from J.A. Baker's *The Peregrine* quoted above, because as I sat making hay while the sun shone between snow showers that last February afternoon, something very unusual was about to happen. Bear in mind that I have watched peregrines on this hillside many, many times over quite a few years now, and while I have witnessed occasional tetchy spats with kestrels (a consequence of which was inevitably some spectacular flying, but never physical injury to either party – a kind of gesture politics), I have never seen anything like the behaviour Mr Baker describes. If you are unfamiliar with *The Peregrine*, it is, by any standards, an extraordinary book. The edition in my bookshelves was published in 2011 (the original in 1967), and features an introduction by Mark Cocker, an accomplished nature writer, naturalist and environmental tutor. He enthuses thus:

Today, it is viewed by many as the gold standard for all nature writing, and in many ways it transcends even this species of praise. A case could be made for its greatness by the standards of any literary genre.

Hmm. In the rarefied world of nature writing in particular, the reputation of Mr Baker's book is almost legendary, so much so that I tend to wonder if many of those who cite it have actually read it or just swallowed the legend. But anyone (especially another nature writer) with contrary opinions would do well to think twice before committing them to print... There, I've thought twice. I believe the reputation is inflated. In places, the writing is undeniably vivid, isolated pools of beautiful prose. But in other places, it's just plain unbelievable, and many people accustomed to witnessing the lives of peregrines have let it be known that they suspect there is an element of fantasy at work. I sail with the doubters in this.

The insinuation that the peregrine consciously mimics the hovering flight of kestrels and the low, slow ground-quartering of a short-eared owl takes some swallowing. Just because the peregrine is famed for its speed does not mean that it is incapable of resorting to one of the many tricks of flight when circumstance calls for something other than speed. It does not need to be shown how to do these things by other species.

And the end of the book, in which Mr Baker disturbs a sleeping peregrine and it doesn't fly away, is surely a dream the author had, and nothing to do with ornithology. I have no problem with an imaginative approach to nature writing. I do have a problem with rendering the imaginative as if it were nature itself. But waiting in the wings, a bird of a very different feather was about to challenge some of my reservations about Mr Baker and his book.

◎ ◎ ◎

So it was the last day of February 2018, the snow fell in frail fragments of showers that brought to mind the blown seed of rosebay willowherb – nothing more robust than that – and I had gone out to the hill to see how far advanced the nesting season preparations of falcon and hawk might be, or if they had even begun at all.

My favourite approach is to walk in the shade of the big trees along the base of the hill until I have reached a point from which I can see the peregrine crag, the gully (the ravens' home) and a certain whin-shrouded rock above a hefty buttress where the kestrels nest. There I like to pause and scan the lie of the land. Between the buttress in the west and the edge of the high crag in the east – about half a mile of hillside – nature contrives a stage set on which it improvises an unending production that never fails its one-man audience and occasionally enthrals. It's like *The Mousetrap* – it just goes on and on. The buzzards routinely drift in from both east and west, not exactly natives of the crag, but they live within sight and sound, and they exploit its potential of lesser birds, rabbits, mice and voles. This is my wide-angle view, where I take the temperature of the hour. Settle and still. This is the best part of the job, re-examining that which is familiar – known landscapes, knowing what to expect, having expectations first confirmed then confounded by something new.

So a male kestrel stepped up and out from an old ash tree rooted in a scree slope, side-slipped across the air and flew parallel to the face of the hill, leading with the open primary feathers of its left wing, then eased to a hovering stand-still above a small clearing of steep, grassy hillside between advancing jaws of whin bushes. One sunlit moment, wings high, tail fanned wide almost vertically below his head, a glowing triangle of pale and dark tawny wings and breast, grey skullcap, black edging to pale silvery-grey tail, sleek in his breeding prime.

Head down. What does he see with his supremely evolved falcon eyes that can differentiate, from fifty feet above, between a blade of grass moving in the wind and the whiskers of a crouching vole a-twitch with fear? It's easy to decide he has eyes for only voles, mice, shrews, small packages of food wrapped in fur. Or – if necessary – beetles, insects, worms. His birding preferences – skylark, meadow pipit, for example – are elsewhere, on the far grassy, heathery flank of the hill.

But might he read the land too? Might he ponder its

changing face? Every spring the whins advance, consume more open ground, meaning less room for voles and mice and worms, less hunting terrain for the kestrel tribe. Arguably it also means more nesting cover for small birds, but that suits the peregrine and sparrowhawk better than the kestrel, for the kestrel is a mouser by instinct, a birder only by necessity, so every spring this hillside becomes a little less kestrel-friendly. Was that, I wondered, in his mind too, forby the need to find the next vole?

I don't have a falcon's eyes, or a falcon's brain, so I see things differently, reason things differently, and I am ever wary of second-guessing what goes on in the animal mind. But from where I sit on the stone wall in the lee of the big trees at the bottom of the hill, and looking up at the kestrel looking down, the bird's dilemma as I see it is this. The kestrel pair has a good nest site on a flat rock atop a small rockface and overhung (and therefore shaded and mostly hidden) by one more whin bush. It has served them well, proved fruitful. But how much longer before it becomes a nest site on a flat rock atop a small rock face in a dark green and bright yellow desert? So do the birds stay put, and if so, must they change their instinctive behaviour to accommodate the changed circumstances? Must they become birders for a living, become more efficient at catching prey in the air rather than on the ground? Or, must they accept that their preferences mean hunting further from the nest and having to carry prey further back to the nest, which means in turn (if Mr Baker is right and the peregrines become as intolerant of nesting season kestrels as he says they are) that they become more vulnerable for longer to falcon-on-falcon predation?

Or do the kestrels pull out?

Do they abandon the safe nest where generations of their tribe have prospered (thirty years' worth to my certain knowledge), take their chances further along the hillside beyond the orbit of this pair of peregrines at least, and beyond the whins, or on the far side of the hill where there is plenty of open grassland but where nest sites could be a problem (for there are no crags and precious few trees)? And will they reason all that out and make a decision, or will instinct simply kick in and evict them? Will nature cry "Enough!" and urge them on their way?

With my questions still poised – kestrel-like – on the hill air, the hovering bird fell thirty feet to a few inches above the whins and sped in level flight towards the blunt bulk of the buttress, a manoeuvre so sudden and low-slung and flat-out that it had the air of drastic evasion. The source of its discomfort was not hard to pin down. The courtship flight of a pair of peregrine falcons is a berserk chase through the heart of the home territory, and as remarkable for its control and its precision as for its raw speed. God knows what it does to the hearts and minds of the lesser fowl of that hillside. Its objective may be nothing more than pair-bonding, but when it free-falls from several hundred feet above the summit, then crackles along the face of the crag from end to end in a blur, many a small furred or feathered heart must skip a beat, many a small beast or bird must freeze in fear or dive for the deepest cover it knows. Given that one of those flimsiest of snow showers was falling and lent a further frisson of subtle beauty to the spectacle, the whole event was simply a thing of wonder in the eyes of a spell-bound nature writer.

The kestrel, unimpressed as far as I know with the concept of a silkscreen of falling snow as a thing of beauty, and given that it knows it can hold its own in the peregrine's company more often than not, neither froze nor dived for cover but rather flew in the opposite direction on a lower contour, and I found him perched in another tree below the buttress, having effected that temporary discretion which is the better part of valour. The choice of perch was interesting. It was in the topmost branches, so not taking cover, not deferring, but knowingly seeking out a particularly wide and unobstructed view of the entire crag. A bit of an object lesson slipped into place: sometimes the sheer spectacle of a predatory presence is a weapon in itself, even when the behaviour is essentially domestic rather than intentionally threatening. The kestrel will be well enough versed in the wiles of peregrine flight to know that when it hunts, it hunts alone, and when two of them cross the territory in tandem in the run-up to nesting time, it is typical nesting behaviour, not typical hunting behaviour. Yet still the kestrel flew, and flew to a more useful vantage point.

So he would have a keen eye for what the peregrines did next, for they segued seamlessly into a long and more dignified oval-shaped circuit across the entire hillside at the level of the base of the crag, a circuit that passed (coincidence or intention – who knows?) a few feet above the kestrels' hidden nest ledge. But this time the perched kestrel did not flinch. As the peregrines completed their circuit, they split apart: the female rose to perch near the eyrie ledge, the male drifted slowly up at an angle and back across the face of the crag to a perch in the big gully, no more than three or four yards from the scruffy, twiggy mound of the ravens' nest. So it had buzzed the kestrel rock and door-stepped the sitting ravens in the same manoeuvre; surely a point was being made about dominance of the airspace over the other predator tribes. It did occur to me, however, that it had happened while the backs of the local buzzards were turned, and if any of the neighbours were built to out-muscle a peregrine...

The thought remained uncompleted, for I still had the glasses on the perched peregrine in the gully when it took off in a manner that suggested urgency, almost a vertical leap, then a steep climb up the cramped confines of the gully's airspace until it cleared the hillside and emerged against the sky. There it began circling slowly, crying out a shrill and repetitive three-syllable note with more than a hint of alarm to it. A second bird crept into the glasses much higher than the peregrine. There followed a few seconds of confusion while I tried to rationalise scale and distance and decided that I was seeing something much larger and much further away. Then it banked, unfurled the full scale of its vast acreage of wingspan and its tail blazed vivid white in the sunlight, and these are the unmistakable emblems of a sea eagle. That frail snow still fell on a light but ice-edged wind from the east, sunlight illuminated it from the southwest so that it glittered as it fell, but the eagle descended from a clear blue northern sky towards the south, towards the peregrine, towards me. I thought (and I may have said it out loud to no one): "This should be good."

⊙ ⊙ ⊙

These are Lowland hills, grassy and plateau-topped, barely reaching 2,000 feet. The crag is at the south-west corner of the range and is an untypical feature. To the south, low-lying flat fields sprawl away to the unspooling River Forth as it heads for its estuary. If you think of the sea eagle as a bird of the Hebridean West, with all of that land-and-seascape's grand gestures (the sheer scope of that repertoire of islands, the illimitable ocean and sky, the surf-washed phenomenon of the Skye Cuillin), the sight of such a bird transported into this comparatively domestic place can be stupefying. It just looks plain wrong, like a whale in a pond.

The eagle descended slowly above the peregrine, which in turn descended slowly above the gully, apparently keeping a healthy 100 feet between them, presumably backing himself to outsprint the eagle if it came to that. I acknowledge an uncharitable moment when I hoped that it might come to that, if only to see what it looked like. But when the peregrine drifted into the open throat of the gully and paused there to execute two level sunwise circles, awaiting developments, the sea eagle did not follow it down. Instead it banked again (another vivid flourish of that tail the shade of sunlit snow) and flew away east along the front of the hills just above the top of the crag and towards the peregrine eyrie ledge. The peregrine's response was to follow, fifty yards behind and fifty feet lower. Both birds flew slowly. The sea eagle's oversized wingspan flapped loosely, glided a few yards with the wings held still but in a shape rather like a wave, flapped again, the whole technique looking as if the bird was making heavy weather of the simplest of level flights. And then there was perhaps the most remarkable moment of the whole encounter between the two birds: the peregrine started to mimic the flight pattern of the eagle, down to the loose-limbed nature of the downstrokes, and although it clearly attempted the wavy-wing glides too, it was not quite so convincing.

"Some peregrines seem to copy the kestrel's way of hunting," wrote J.A. Baker. And: "…In a similar way, I have seen a peregrine hunting near a short-eared owl, and apparently mimicking its style of flight." Until now I was blithely associating myself with the doubters among Baker's readers, but here was what certainly looked like deliberate mimicry of a sea eagle. But to what end? What on earth is the point? Is it play? Is it a kind of flattery, an attempt at ingratiation? Is it deliberate and calculating or merely instinctive? Baker says "*some* peregrines" as though he thought it was not a universal trait in the species. It occurred to me, too, that if it had been a golden eagle rather than a sea eagle, it would have flown along the top of the crag in a stiff-winged glide without a wingbeat, a feat it can reproduce at almost zero knots, and I suspect that might not be in the peregrine's repertoire. But is that the point, a purposeful exploration of one more of the possibilities of flight beyond the peregrine's renowned capacity for flat-out speed, and perhaps guile is the end product in order to fool predator or prey species, the way a skylark or a dotterel feigns a broken wing? At the very least, I need to re-appraise some of Mr Baker's ideas.

I watched both birds until they vanished beyond the skyline that marks the end of the crag. I wondered how far falcon would shadow eagle. I guessed until it was well beyond the margins of the home territory. I gave him another hour, but he did not reappear. I imagined him sitting upright on a ledge or the top of a conspicuous tree, staring east where the eagle had vanished, lost to even the phenomenal eyesight of a peregrine falcon.

In the meantime, it had stopped snowing.

⊙ ⊙ ⊙

A long spell of wintry weather relented on April 9th, and something recognisably spring-hued crept over the horizon and seeped along the woodland at the foot of the hill. I could almost taste the difference on the air. There were two buzzards and two kestrels above the west edge of the crag. The unmistakable mewing buzzard voices drifted down and rubbed shoulders with the hysterical, giggling falsetto of a green woodpecker. The wind had veered from east to south, the temperature had eased away from the frigid zone and nudged into double figures – just. It was the first day I had seen flying insects, and no sooner had the realisation lodged in my mind than I heard the first chiffchaff, something like three weeks late. By the time I was back at the car four hours later, I had heard six different calling chiffchaffs. Did they all arrive at once? But the yellowing of the hillside, which is such a characteristic of spring here, had stalled halfway up, and even the whin blossom on the lower half was nothing like the density of old springs.

I paused at my accustomed stance against the wall that edges the wood, and the buzzards were pretending to be golden eagles, which is always worth watching if you don't happen to be in golden eagle country. The female was locked into a horizontal spiral that travelled east but seemed not to gain height at all, the opposite process to that which is evident in most buzzard spirals, but if your purpose is to scan the hillside at a leisurely pace for rabbit traffic across the small clearings in the whins, then such a flight technique is perfect for the job in hand. Her mate was much higher and almost stationary, holding up against the wind, working his tail more than his wings, so that from time to time the silhouette was almost kite-shaped, for no bird in these skies leads more fluently with its tail than the red kite. Maybe Baker was onto something and they all borrow from each other when living in such close proximity.

Suddenly the kestrel pair was charging down through the airspace between the stationary male and the sideways-spiralling female, apparently intent on heading off the female's eastward journey, which was now more or less directly over their own nest ledge. My long experience of this hillside concludes that kestrel–buzzard action is much more routine than that of the kestrel–peregrine variety highlighted by Mr Baker, but perhaps his corner of the Essex coast was devoid of buzzards in the 1960s. The male buzzard now tipped forward off his airy perch, half closed his wings, and fell into the fray, and for perhaps half a minute all four birds tore up and across the hillside and in and out of sunlight and shadow; the kestrels chattered and the buzzards mewed, and I was aware of nothing else on the face of the land. No blow was struck, and the whole thing had the air of ritual rather than outright aggression. Then the buzzards eased away round the west edge of the hill, where I have always assumed they nest out of sight from here, and the kestrels perched a few yards apart, one in a fine old ash, the other on a nondescript hillside rock, both perches within yards of their hidden ledge.

The whole encounter had been so blatant, so conspicuous, so demonstrative, that I was surprised that it had not attracted the attention of the peregrines. Or had it? Sometimes, watching nature on this hillside seems to be ridiculously easy: when I turned my glasses towards the peregrines' end of the crag, there was the male on what I think of as his look-out rock, standing erect, and I wondered (a) how long he had been watching, and (b) whether all peregrines have regular look-out perches. There is a corroboration of that idea in James Macdonald Lockhart's book *Raptor* (Fourth Estate, 2016), in the unlikely setting of Coventry Cathedral:

When I reach the bench I find them as I had left them the week before. Both birds are perched above me, the male on a ledge high on the spire of Holy Trinity, the female on the flèche of the new cathedral, directly opposite her mate. The bench sits in the ruins of Coventry's old cathedral. It is a contained space away from the noise of the city, not unlike the corrie on the side of the mountain in Sutherland, a place of enclosed stillness, where the ruin's walls hold and amplify the sound of the falcons calling to each other... One June, whenever I had a spare morning, I hurried back to the cathedral ruins, arriving at the bench soon after dawn, the sun turning the cathedral's sandstone a deep chestnut red... The falcons were always there, at home, often perched in the same place where I had left them.

In the glasses he looked tall, taller than you would expect if you are familiar with the bird in every guise other than standing erect on a look-out rock. It is particularly marked with the larger female. They are not large birds, but they strive to create a larger-than-life presence whenever they perch near the eyrie. The notion disintegrated the instant the perched male flew and resumed his compact size and shape. That first handful of wingbeats always looks clumsy to my eyes, even chaotic, but after a few yards it smoothes itself out into supreme fluency. This one proved to be a short flight, a sharp turn back along the face of the crag and an arrow-straight climb into the rock face, disappearing in the instant of landing. That old trick! I knew exactly where he had perched, the rock was well lit; I have good binoculars, but I couldn't see him. The peregrine has that capacity to become rock if it chooses.

For the moment there was no bird in the sky, and I set off for my own favourite perch in all that landscape. An old rowan tree stands quite alone on the hillside, and for the moment at least it is beyond the reach of the whins. On its uphill side is a perfect grassy couch, and with the trunk for a backrest I can sit there in relative comfort for hours, facing uphill, watching the entire length of the crag, the entire height of the gully, the hillside sprawl below the crag (where a fox or a roe deer might materialise at any time) and a huge tract of sky. I settled, drank coffee, tuned in to the penetrating calls of chiffchaffs that filtered up through the wind from below, their sudden confirmation of spring's arrival.

I was thinking about birds of prey and their nests, or lack of them. If kestrels and peregrines can function perfectly well without building a nest (an occasional compromise is made by squatting in an old crow nest), why do buzzards, sparrowhawks, kites and ospreys need them, and why does a golden eagle build so massively on a mountain ledge or transform the size and shape of a Scots pine, always adding to the previous year's structure? Sea eagles also build massively. Golden eagles, in particular, constantly freshen the nest throughout the nesting season, bringing green sprigs of rowan and birch and weaving them into the structure. Why does a peregrine not do that? Why is it willing to settle for a bare crag like this one, a sea cliff, a window ledge on a city tower block? Or, for that matter, a ruined cathedral?

I spent two agreeable enough hours sitting by the tree, making notes, scanning the crag, the hillside, the sky, but it was as if a kind of torpor had taken possession of the landscape. I had seen no peregrine in all that time, and of all the conspicuous residents of the crag only the ravens had showed themselves. By late afternoon I was back at the foot of the hill, walking slowly by the edge of the wood, reluctant to take my leave of the place. That old familiar sense of unfinished business.

The buzzard pair from the east, beyond the peregrines' ledge, came spiralling across the hill, that other familiar sense of recognition that belongs uniquely (in my experience, at least) to this hillside, and by which they travel across the front of the hill while executing flat spirals, each circuit concluding a little further to the west. So I began exploring in my mind the extent to which any one set of landscape circumstances infiltrates and makes demands on the characteristic behaviour of individual species. These buzzards could traverse the hillside in straight lines and at different levels, and they do, but just as often they choose the flat spiral technique, and by constantly circling they can effectively see what's going on behind them as well as what lies ahead, and as the land falls away beneath them then rises again they scan the uphill and downhill of the view with the minimum of fuss. And if they too have a problem with peregrines, theirs is a technique of travelling that watches all the compass points all the time. It's clever. Is it rationalised, or is it instinct?

I have no way of knowing, but I do know (from a long-term study of swans, as it happens) that there is individuality, of character, within what too many of us are inclined to accept unquestionably – what I think of as "field guide syndrome", the tendency to generalise, to discard the possibility that there are smart and stupid swans, sociable and stand-offish swans, aggressive and passive swans; also that wild creatures often work within a species of logic that bears no resemblance to human logic, so we are capable of completely misunderstanding a particular behavioural trait because we apply human logic to what we see rather than, say, peregrine logic. No job requires a more open mind than mine.

The buzzards stopped abruptly. One flew into the big trees and perched high up. The other flew down to no more than a yard above the whins, picked up speed, drew in its wings very much in the style of a golden eagle. Over a hundred yards of hillside, it flew in and out of the whins, then suddenly rose at a shallow angle towards a big ash tree less than fifty yards directly uphill from where I stood. At once, a peregrine flew out of the tree with that ungainly shimmy while the wings co-ordinate, then a superbly sleek and low-slung burst of exquisite speed beneath the buzzard and away across the hill. The buzzard perched in the tree the peregrine had just vacated. What it looked like was an eviction, an assertion of a pecking order, of hierarchy. That's what it looked like. I have no idea if that's what it was. I know of peregrine enthusiasts whose observations include buzzards suffering under a peregrine regime. In *Raptor*, James Macdonald Lockhart writes:

Almost all birds are at risk from peregrines. There are records of geese, black-backed gulls, even buzzards being struck down.

Yes, there are. But are those anomalies that have been folded into the peregrine lexicon by those who have a vested interest in assembling a peregrine mythology? Again, I don't know, but I think there is a good chance that, as with Mr Baker's *The Peregrine*, the bird's role as a kind of superhero of nature is overstated. I had just seen a buzzard respond to the sight of a peregrine perched in a tree (a position from which it could just as easily have launched an attack on the buzzard if it had been in the mood) by taking a singularly aggressive initiative. Seconds later, the buzzard's mate arrived and perched in the ash, and after a few more seconds, they mated there. Thinking about it all now, I wonder if the outcome had been determined by a simple calculation that was self-evident to all parties: it was two against one.

I found the peregrine in my glasses again when it broke the skyline and dropped to a high, wide-open, level, grassy terrace, the grass still bleached to that fag-end-of-winter shade of no recognisable colour at all. It perched there, and as it did so, another bird I had simply not seen broke cover, flew the length of the terrace at almost ground level, and pointed straight at the perched falcon, which once again took flight. This new bird flashed an emblematic tribal badge no less conspicuous than the sea eagle's tail, a white blaze at the base of a long-straight tail. It was a female hen harrier. This just wasn't the peregrine's day, but it made mine.

But that day was not quite done with me yet. The peregrine vanished behind me but the harrier wheeled and retraced its journey along the terrace, then alighted on the ground. There, after fussing with something in the grass, she rose a yard above the ground, wheeled once more, and traversed the ledge for the third time, but this time her attitude was quite different. She travelled at the speed of a weary carthorse after a long day between the shafts, and her sleek profile was troubled by a hefty morsel of prey bunched up, dark and bloodied, and stashed in her talons beneath her. I don't know what it was. I guessed that – wittingly or otherwise – the peregrine had disturbed the harrier feeding in the grass, and that the harrier had retreated rather than defend her prey, possibly as a matter of tactics rather than fear. But the peregrine did not attempt to steal the unguarded prey, so once again I was questioning Mr Baker's book. Ah, but just at that moment, as the harrier inched low and slow along the terrace, the peregrine returned, not to attack the harrier but to mimic its flight from about fifty yards behind it, keeping a distance, showing no aggression, but perhaps – just perhaps – escorting her off the premises.

So I am no closer to drawing reliable conclusions about Mr Baker's book, and as he died in 1986, it's too late to ask him about it now. I came to it very late. I came to all English nature writing late, for I was raised on the voices of my own land – Burns, of course, and perhaps he was the first nature writer of all of us, Seton Gordon, David Stephen, Tom Weir, Don and Bridget MacCaskill, Adam Watson, Nan Shepherd, Gavin Maxwell, Mike Tomkies, and a clutch of poets that included Norman MacCaig. With so many voices steeped in the landscapes in which I labour myself, what need did I have of England's scribes? Ah, but then, of course, I found the writing of Margiad Evans, whose soft-voiced and close-focused nature writing took my breath away, and I went from there, and it was on the journey of discovery which followed that I met J.A. Baker's radically different book about peregrine falcons.

The falcons of the yellow hill had to endure – like so much else in the spring of 2018 – some very untypical conditions, not least of which was that the yellow hill never did turn that brilliant top-to-toe yellow that is how I had come to think of it. A lot of things changed that spring, and one of the consequences of writing it down will be to remind myself to re-read and re-appraise *The Peregrine* by J.A. Baker.

Chapter Three

The Backward Spring

A Backward Spring

The trees are afraid to put forth buds,
And there is timidity in the grass;
The plots lie gray where gouged by spuds,
     And whether next week will pass
Free of sly sour winds is the fret of each bush
     Of barberry waiting to bloom.

Yet the snowdrop's face betrays no gloom,
And the primrose pants in its heedless push,
Though the myrtle asks if it's worth the fight
     This year with frost and rime
     To venture one more time
On delicate leaves and buttons of white
From the selfsame bough as at last year's prime,
And never to ruminate on or remember
What happened to it in mid-December.

Thomas Hardy,
April 1917

ONE HUNDRED AND ONE years after Thomas Hardy's pen
shivered its way through those lines, having fun with rhyme

schemes even while he was making a serious point about the jeopardies of facing nature when spring sleeps in, the spring of 2018 was posted missing. On the very day that the Met Office announced:"Tomorrow is the first day of meteorological spring," what would turn out to be a foot and a half of snow began to accumulate in the garden, changing its shape, obliterating its short flight of stone steps, blurring the distinctions between lawn and rockery and shrubbery, and cloaking everything in bloated silence.

Two things happened that gave me pause for thought. The first was that my one dependable symbol of the end of winter (an early-flowering dwarf rhododendron whose native airt is Himalayan) gave up the ghost. Every year for forty years, its flowers have preceded its new leaves, and it has illuminated February-into-March days with a low-lying cloud of pale pink flowers as cheerful as it is incongruous. Even after what had already been a cold winter, it began to bud in the accustomed manner in February of 2018, then the snow arrived on Siberian easterlies, and once the snow melted again after about five days, what was left was shrivelled brown. It took until mid-May for its new leaves to recover, but there were no flowers.

The second symbol was at the end of the street where an open area of grass and trees is bordered by a wee burn, an innocuous-looking thing that spends much of its life underground, but it played a decisive role in the Battle of Bannockburn (it is a tributary of the Bannock Burn and apparently helped to screw up any meaningful progress by the English cavalry, back in 1314; there are long memories hereabouts). A few years ago, on the 700th anniversary (a good excuse for a party, given those long memories), the local council planted early-flowering daffodils along the top of the bank. But in the February of 2018, they abandoned their early-flowering tendency and gave up on the idea of being daffodils for a full month. I should explain, perhaps, that cultivated flowers are not my thing; I like wild orchids on Lismore, starry saxifrages on Highland mountains, twinflowers in pinewoods. Gardening and I hold each other at mutually agreed distance and regard each other with mutually agreed distrust. But those daffodils and that rhododendron – these two standard-bearers on the frontier between winter and spring – are exceptions to the rule for which I graciously concede my gratitude to horticulture.

Springlessness, then, was suddenly rife across the length and breadth of the land and making news and causing more or less wholly unnecessary chaos to society as we know it. Climate change deniers stirred themselves in their dinosaur lairs, dipped their pens in bile and gloated in the letters pages of complicit newspapers, braying variations on a theme of "whatever happened to global warming?" But, predictably, they resumed their stupors and their slumbers when the whole thing was over in five days, and within a couple of weeks London and the south-east recorded temperatures in the twenties, then within another couple of weeks, in single figures. Climate chaos is what we have, what we are responsible for and what we suffer from. And as the backward spring cowered in its chrysalis long past the moment when we might reasonably have expected the butterfly to flex its peacock wings, and even as Siberian winds whipped fallen powder snow into drifts the depth of breakers in a surfer's paradise, spring in the real Arctic was arriving two weeks early. Polar ecologists at the University of California, Davis, introduced newly published research in terms that gave me a third pause for thought:

Spring is arriving earlier, and the Arctic is experiencing greater advances of spring than lower latitudes. Over the past ten years, spring has come a day earlier, but two weeks earlier in the Arctic, where temperatures are rising twice as fast as the global average and ice fields are rapidly shrinking.

The report added that in April 2018 temperatures in the High Arctic (where there had been no sunlight since the previous October) had been above freezing level for a total of sixty-one hours. Then, as May dawned in Scotland with raw winds, and a dearth of both migrant birds and insect life for them to feed on, a team of British and American scientists drew the world's attention to the changing condition of the Thwaites Glacier in Antarctica, "a structure that drains an area the size of Britain". The change suggested potential for rapid melting or even complete collapse. So far, changes to the glacier have only been detected by satellites. The scientists were announcing the beginning of a five-year project to examine the glacier at close quarters, to find out exactly what is happening. If their worst fears are realised, the potential for global disaster is more or less limitless. The scientists encapsulated the difficulties ahead thus:

We do not know how quickly the glacier will contribute to sea level rises, and whether we have decades or centuries to prepare for it.

Not "to prevent it", please note, but "to prepare for it". One risk is global inundation of coasts – and therefore coastal cities, towns and villages – because if things are going horribly wrong with the Thwaites Glacier, one consequence could be a rise in sea levels of 1.5 metres. To put that into context, the Intergovernmental Panel on Climate Change (established in 1988 by the United Nations Environmental Programme and the World Meteorological Organisation, and with a membership of 195 countries) believes that if humankind embarked on radical and sustained action on greenhouse gases, we *could* limit sea level rises to something between twenty-five and forty centimetres, but that does not allow for an event like the collapse of the Thwaites Glacier. The bill for the Anglo-American Thwaites project is £20million. If it averts or even moderates global consequences, it will be cheap at twenty times the price. But meanwhile, be afraid.

One more symbol of the backwardness of spring 2018 came from the Woodland Trust. The trust runs a project called Nature's Calendar, in which members send in details of the dates when significant events in nature's year actually occur, in this case, the first records of bluebells – or wild hyacinths – in flower. In 2017, the first recorded date was in the south-west of England – February 9th. In 2018, the first recorded date was in the south-east of England – March 20th, or thirty-nine days later.

By April 20th 2017, the trust had received 716 records of flowering bluebells. By the same date in 2018, there were seventy-three.

⊙ ⊙ ⊙

March comes in like a lion and goes out like a lamb. It is one of those clichés with which the folklore of our seasons is carelessly strewn. The missing word is "sometimes". In the backward spring of 2018, March came in like a polar bear and went out like a flu victim, spluttering and out-of-sorts. For more than thirty years, my idea of spring was articulated by the return of sand martins to a certain bay on the north shore of a certain loch, because for more than thirty years I had followed the fortunes and misfortunes of a pair of mute swans that nested in a reed bed that rounded off the north edge of the bay. I would keep tabs on the swans throughout the winter, but once the spring set in, I settled into a more organised routine of watching them prepare for the nesting season, a process which, among many other characteristic events, involved clearing out any unwanted birds and beasts from a large area of the north end of the loch and a lochan beyond it.

And sooner or later, and more or less reliably in the last two weeks of March, I would arrive at the loch to find the first few feet of airspace above the surface aswarm with the dull brown charms of hundreds of sand martins. "More or less" because I was never one to keep records of such things; I own nothing remotely like an archive, and I have always been content with the knowledge that they have returned, and to acknowledge it with welcome and gratitude. Repeatedly, and over those decades, the sand martins slipped unobtrusively and on cue into nature's rolling scheme of things on the loch, the swans' scheme of things too. And always, within days of their arrival, the flock would fragment and a hard core would drift a mile upstream from the loch and settle in where a ten-feet-high bank of bare earth curved round the inside of a long bend in the short life of the River Balvaig.

The Balvaig flows east out of Balquhidder Glen, coils and uncoils lazily across a wide, flat-bottomed floodplain known locally (but not on any maps) as Loch Occasional, beyond which (still coiling and uncoiling) it heads more decisively south to Loch Lubnaig, where its long, tree-girt thrust into the loch creates two bays, one of which is the undisputed realm of the swans. And halfway between the two lochs the river heaves ponderously from east-making to south-making by way of one long bend, and it is there that the bank rises to ten feet above the water, there that the sand martins call home, and there that they begin with formidable industry to excavate old and new burrows in the bank's accommodating soil. The first sites to be claimed are always the highest, the ones sheltered by a grassy overhang, a green cornice. It is there, too, that the greatest concentration of nest burrows lies, for they are also safest from sudden spates and rising river levels, and the Balvaig is particularly accomplished when it comes to dismissing the strictures of its banks.

The backward spring produced headlines all across the country proclaiming that nature's migrating tribes were far behind schedule and, having confirmed notable absentees from within my own workaday landscapes, it was the second half of April before I contemplated an expedition to check on the martins' progress. The path from road to riverbank runs along the top of a long, narrow strip of woodland that falls steeply to the "shore" of Loch Occasional. The wood is an open one of oak, birch and willow, with a lush, grassy and ferny understorey. The river itself only comes into view when the path emerges beyond the end of the wood, and from there a short diversion between small fields of rough grazing leads to the bank directly opposite the martins' headquarters.

A morning's anticipation of reunion dissolved in about five seconds. Not only were there no sand martins, there was no river. The vast tonnage of snow that had assembled on Balquhidder Glen's surrounding mountains and lesser hills was responding to the most grudging of thaws by pouring down mountainsides in avalanches and impromptu water-falls. Loch Occasional was a square mile of shallow water covering a treachery of grassland and bog, but also includ-ing much of the watercourse of the river itself. Its presence was only traceable by curving lines of trees in single file where they still clung tremulously to submerged riverbanks. To hosts of small birds, up to and including waders, and to mammals of any size, shape, or persuasion, up to and includ-ing me, the place was simply out of bounds, for nothing about it was reliable. On the other hand, if you were a swan, a goose, or a duck, Loch Occasional was a part-time para-dise, for nothing suits the web-footed, long-necked tribes better than being able to feed on grassland through shal-low water. A slow sweep of binoculars revealed mute swans and whooper swans, greylag and Canada geese, mallard, goldeneye, goosander, wigeon, teal, tufted ducks and little grebes. And alone among that host, a widely scattered presence of solitary herons tried to make sense of their temporarily redefined world.

The following day, I tried a little nearer home, where a much smaller but no less reliable sand martin colony on the Sheriffmuir edge of the western Ochil Hills was equally bereft. But an area of grassy mounds and grass-and-heather moorland nearby revealed prolific burrowing, not by martins but by short-tailed field voles, and there were thousands of new holes. Such profusion lures predators, and the droppings of fox and pine marten were equally profuse. Likewise, a flat-topped rock that I know to be a regular short-eared owl perch had pellets around its base like pebbles on a beach. And that day I heard the year's first skylarks and, laving their columns of song with jazzy woodwind licks, three curlew voices confirmed that nature had not completely forgotten its repertoire of spring anthems. A red kite inched uphill into the narrowing confines of a handsome little glen, the bird's rich colours deepened by the dourness of the day, the poor light, the imminence of rain, and it too was vole-hunting.

Then I turned away west to face where the mountains of the first of the Highlands were lost behind an advancing frontier of new rains. There was an extraordinary beauty in that sky that fell unbroken to the nearest edge of the moor, for it was horizontally bisected high up, and while all was dark and glowering beneath that line, above it was a much paler shade, just as uniform as the other, but somehow refreshing, even invigorating. I could imagine Mark Rothko standing where

I stood, contemplating that sky and its possibilities. For in my mind was a painting from 1963 where two almost mono-chrome bands of such shades were kept apart only by the slimmest dead-straight horizontal line of white paint. It's in a copy of the catalogue for a memorial exhibition in Venice following his death in 1970, and which was given to me by my great friend George Garson, one-time head of murals and stained glass at Glasgow School of Art. He had gone to Venice to see the exhibition. Just inside the front cover of the catalogue, on a page by itself, is the following:

Telegram received by one of the Executors of the Estate of Mark Rothko, deceased February 25th, 1970:

"Profoundly moved by Mark Rothko's tragic death. Remember in deep sorrow his voice, his gestures, his very secret and enlightening radiance.

His painting is one of the most magnificent in this century and leads to a new wholeness of thought and vision.

Franz Meyer, director, Kunstmuseum, Basel."

"A new wholeness of thought and vision"…now there's an epitaph worthy of the occasion. And a catalogue repre-sentation about four inches wide of a canvas that was ninety inches wide and painted fifty-five years before I stood that day on that moorland edge in the backward spring of 2018 looking for sand martins, had just stopped me in my tracks and made me appraise a grey sky in terms of great art, a sky that many people standing in my shoes would have described (if you had asked them) as "dreich".

I shook my head in barely comprehending wonder and thanked the man silently for his art, for his new whole-ness of thought and vision, and I thanked (for perhaps the thousandth time) my dear-departed friend for handing me the catalogue one day in 1991 and saying something like: "You're ready for this now." I'm not sure. Even now.

Then the moment transformed. The three curlews lifted unhurriedly from the moor and cruised into Rothko's sky, drenching it in guilt-edged threnody, then descended in silence and vanished below the skyline. It was just as well, perhaps, that I was on my own right then, for no one would have got a word of sense out of me for the next hour.

◎ ◎ ◎

The martins came eventually, but in half-hearted numbers, and it was difficult to see how they could retrieve much from such a backward spring. Loch Occasional did what it always does and shrank back into the floodplain; the river did what it always does and adapted its girth and depth to fit snugly back within its banks, revealing that the sand martins' bank had survived one more inundation. I kept a discreet distance from the bank, watched just long enough from the cover of trees to confirm that those martins that had bothered to complete the journey were doing what they should be doing, then I turned back. Whatever the prospects for them now, my pres-ence would not help. Instead, I lingered the morning away at the edge of the wood, where the high-water mark of Loch Occasional would have been less than a fortnight before.

There were two swans far out on the floodplain, feeding among its myriad pools. I thought at first they might have been the mute swan pair up from Loch Lubnaig, taking advantage of easy feeding. But once I settled binoculars on them I realised they were whooper swans, and in a more forward spring than this one, whooper swans would have been back in Iceland by now. I was attracted to the idea that they might stay, of course, and try and nest on one of the glen's two lochs or by the river. Such things do happen every now and again, the swans finding enough of an Icelandic atmosphere to persuade them to stay. It happens especially in the far north of the Scottish mainland and in both the Western Isles and the Northern Isles. But there are relatively recent precedents, too, in places like Loch Lomond, and in Highland Perthshire where a pair nested for three years but had their eggs stolen each time, and they finally departed, suitably discouraged, as well they might.

Ornithologist Mark Brazil's monograph, *The Whooper Swan* (Poyser, 2003), notes that "whoopers have a long and traceable history in the British Isles", that as glaciation ended and ice sheets shrank northwards, the tundra-like aftermath would have been perfect for them, and there is evidence to suggest they bred regularly during the Little Ice Age between the 15th and 19th centuries. Then, after a long absence in the 20th century, whoopers have been breeding again since 1970. In 2003, the bird was considered to be "a rare and erratic breeder". That still sounds about right, but if the climate continues to warm, and our winters grow generally milder, the birds may vanish from our shores once more, taking other Arctic migrants back with them.

Meanwhile, by early May, there were still those two birds scaring the resident Canada geese witless and adding their bright white grace to the landscape, and consoling this nature writer and unashamed fan of whooper swans on one more troubled morning of the backward spring. Occasionally a muted-brass syllable of their conversation drifted across to me and mingled agreeably with the year's first snipe voices. And on the hillside across Balquhidder Glen, I could see a handful of red deer out on the open hill feeding in pale sunlight. What began to unfold was one of those situations in which I delight: very slowly, my absorption in my surroundings and the small movements of nature across the low ground became a moment, a blink of time in the testimony of the mass of mountains beyond. I was of no more account than that snipe calling metronomically on a tussock, arguably of less account than that pair of Icelandic itinerants or the incomer Canada geese, or the ancient, seen-it-all-before heron up to its shins in tepid bog water. And for the purposes of all these creatures, I had become landscape. For a nature writer, it is in circumstances like these that anything can happen.

How long I sat on a moss-draped rock with my back to an alder trunk before I realised it was already happening above and behind and over my right shoulder, I have no idea. But I became aware, as if from a great distance, that a new sound was abroad on the land, not out on Loch Occasional, not on the hills or the river, but in the wood that surrounded me to north and south and east. I believe I had registered the sound and that it was getting closer before it began to make enough of an impact on my mind, my awareness, for me to respond to it. Such moments exist in a different kind of time to the one I use to measure my days and nights, and it passes more slowly and more meticulously, and more preciously, too, for it permits me to respond to nature's presence both as a whole and in its component parts. The eternal problem is that my response is never capable of holding on to it all, far less write it down, for I am constrained by the writer's limitations and my subject is limitlessness.

Finally, the sound came clearly into focus. I have been through situations like this many times now, and I have taught myself that if movement is necessary, then it should be very, very slow. I am, after all, trying to be a bit of the landscape. This much I decided at once: something was moving through the wood across its slope from my right to my left so that soon it would pass behind my back. It was moving slowly and stopping often, and when it stopped there was a strange rasping sound, with the rhythm of a handsaw being deployed in short bursts. It was that sound I had been hearing at the edge of my consciousness while my mind was elsewhere.

The mossy stone where I sat was only about a foot high. The tree trunk at my back would screen most of me. I slid forward off the rock as slowly as I could, lowered my upper body and rolled over in the same moment so that I was prone on the bottom of the slope and facing uphill. I had done it as gently and quietly and fluently as I knew how, but when I raised my eyes, it was to meet head-on the full-bore, double-barrelled stare of a roebuck at ten yards. He was thickly cloaked in his grey winter coat. He had wintered well, he looked solid and muscled and well-fed. His six-point antlers were newly cleaned of velvet and smeared in blood by the process. He lowered his head even as I watched and slashed them several times left and right among the grass, ferns and brambles, a gesture that merged elements of brute strength, bad temper and natural grace. Whether it eased the irritation in the new antlers or provided an outlet for pent-up breeding season energies, I could not say, but when he raised his head again and stared hard at me, it was hard to resist the notion that hostility lay behind the eyes. It could not have been helped by a broken length of bramble bush that had become snared in the tines and hung down in front of his eyes. He lowered his head again, raised a front foot to pin the bramble to the ground and jerked his head upwards to free the obstacle. It was so deftly done that it was surely a well-practised routine.

He turned his head sideways and barked twice, and I saw his breath hang in a cloud in the morning's cold air. An answering bark came from the trees not twenty yards away, and at once the buck and his unseen doe were running through the trees, their rhythmic barking underscored by the sound of their feet. The next time I saw them they were far out on the floodplain heading towards the river, and they took with them the shreds of the frail binding that had held together those moments of enchantment in which I had briefly lingered.

⊙ ⊙ ⊙

One of the by-products of writing this series of books is that I have started to scrutinise the performance of all the seasons of all the years of the project, which in turn has sharpened my focus on climate and its effect on what most of us consider to be an annual seasonal pattern in which every season has a function to perform. A particular sense of disquiet set in early. I was vaguely aware of it as a kind of recognisable passing stranger with whom I had never troubled to communicate. Urged on by nature, the disquiet has since troubled to communicate with me. It has already infiltrated both of this book's predecessors, *The Nature of Autumn* and *The Nature of Winter.*

Arguably, its seed was sown in 2015 when the Met Office decided witlessly to call a particularly troublesome storm "Abigail", thus launching a new policy of calling storms by their Christian names, and imbuing them with a sense of smirking, domestic, Disney-esque gaucherie that somehow belittled their true nature as phenomena that gathered succour and oomph from globe-trotting oceanic voyages. But when Abigail berated our landscape with a formidable tongue-lashing, its message was not delivered in a Scottish accent but in global speech, a kind of meteo-rological Esperanto. In *The Nature of Autumn,* I wrote about how that disquiet's first impact was easily identifiable as a palpable restlessness of nature in the heart of my everyday landscape, and how it had struck me in the same moment as part of something much bigger. So I had made a space in a particular landscape I know very well in order to listen to the land.

And in *The Nature of Winter,* from the midst of a winter in which the only traditional winter characteristic to be found was an absence of daylight, I allied that disquiet to the announcement that a huge glacier in northern Greenland called Zacharie Isstrom was found to be melting at the rate of five billion tons a year. Now, from the midst of a spring in which the only traditional spring characteristic to be found (at least for its first two months) was lengthening daylight, winter has not so much clung to its coat tails as ripped the tails from the coat and buried them under depths of snow; just in case the announcement about the Thwaites Glacier in Antarctica had escaped our notice.

The disquiet grows, intensifies, becomes clamorous. The land is talking to us again. For nature's sake, for the Earth's sake, and yes for our own sake, for God's sake, listen.

⊙ ⊙ ⊙

Admittedly, it is getting harder to listen. Also admittedly, this may be more of a problem for a nature writer than most people, but the careless volume at which everyday human life is pitched has become troublesome to me. Never has quietude seemed so elusive. If it were a species, it would be on the verge of extinction. And you cannot launch captive breeding programmes for quietude.

An estate agent's for sale ad described a neighbouring house as being in "a pleasurably peaceful edge-of-town cul-de-sac". Whoever wrote it, they were not here when the gas pipes were being renewed just after the tarmac had been re-laid, when the extensions were being built, and sundry domestic and local authority and builders' leaf blowers and power saws were singing along to the strains of two different radios tuned to two different stations. Every vehicle attending to such enterprises had to reverse the length of the street because the enterprises rendered the turning area not negotiable, and as they reversed they engaged the "I am reversing" bleeper, a species of noise pollution designed to communicate the bleeding obvious. "Scottish Gas – looking after your world" was emblazoned on several vans. Not my world, it isn't.

After one particularly worse-than-fruitless morning, I took evasive action, packed a water bottle, a couple of bananas and a notebook and pen into a small backpack that already had two cameras, a mat and a folded bivvy bag to sit on, as well as a compass and some chocolate as permanent fixtures. Forty minutes later I had parked my car in some trees, and fifteen minutes after that I stepped downhill from a well-walked path into the complicit depths of an oakwood. The hyacinth-scent of bluebells rose up to meet me on a warm draught, and it was accompanied from further down the hill by the voice of a river concealed from the bluebell wood by a sprawl of rhododendrons. I tiptoed down through the patches of blue that were darned and hemmed with bright-white stitchwort, I bashed a none-too-considerate way through the rhodies, I squelched around the rim of a small bog, taking advantage of some rotting broken branches that reduced the impact of the glaur, and I emerged onto a sunlit rocky corner of riverbank, and there the healing balm of riversong was a beauteous mercy to my every sensibility. I sat. I breathed deeply. I looked round. This is my idea of home, not that otherness I had just walked out on. This is where my sense of belonging lives; not necessarily this particular river, but the embrace of nature symphonic in my ears and my heart and mind. I took out the notebook and pen and began my working day again in the middle of the afternoon.

If you travelled towards the first of the Highland mountains heading north-west from Edinburgh or Stirling, the
first sight and sound and music of a proper Highland river
would be right here. I perched on the rocks six feet above
the river and the same distance from the edge. The river was
deep and tranquil immediately below me, shallower at the
far side. There were rocks enough to produce white water
and gentle waterfalls fifty yards upstream and thirty yards
downstream. The upstream white water glittered silver in
full sunlight. It emerged from beyond a bend in the river
and it was a band of fractured white from bank to bank. But
the river grew calm almost at once, and by the time it had
reached my rock it was all but silent, except that a tiny riffle
half an inch high and a foot long bubbled a short thread
of magic and a song pitched right on the very edge of my
hearing. If I looked at it directly, though, that helped, and
the tiny dancing of the thing laid a detectable vibrato on its
essentially contralto whisper; sight exquisitely and minutely
enhanced the sound.

Of the two waterfall voices, the upstream one was both
louder and higher-pitched, and its rockier course was more
percussive. The inside of the bend produced the throaty
notes; an elegant little fall hard by the outside had a musi-
cally metallic, xylophonic edge; and the widest stretch of
white water between these two had ordered its twenty-or-so
mini-falls into a fluent blend from which the whole fashioned its essential, recognisable sound. The whole time I sat
there – about three hours, at a guess – there were birds singing from the nearest trees (some overhanging, some whose
roots waded into the river margins). At different times, these

included wood warbler, willow warbler, chaffinch, dipper, blackbird, song thrush, goldfinch and one-woodwind-note bullfinch, and I wondered after a while whether there was anything in the idea that riversong stimulated birdsong.

Mostly, I sat facing upstream, partly to have the sun on my face, partly because upstream was the more notable fall (yes, the prettier one), partly because the rock where I sat was more accommodating of an upstream-facing sitter, and partly for the view. That band of sun-silvered white water was backed by a row of ten tall conifer trees – Scots pines and Douglas firs – and a mass of birch and oak in vibrant spring greens, and beyond and above all that, the handsome blue-green bulk of the summit and north shoulder of Ben Ledi was draped across an all-but-cloudless sky. No mountain in this part of the world dignifies its landscape setting with more natural grace than that one. My eyes and my heart and everything within me that contributes to my nature writing instincts are gladdened by its presence more days of any one year than not; I celebrate it, I am gratified by it, and I greet and leave its presence with a glad hand, and my mind mutters to itself: "here's a hand my trusty fiere…" for you can sing that immortal slice of the genius of Robert Burns to greet an old landscape as well as a new year.

The downstream fall had no such backdrop and no singing birds. But from where I sat, it provided an agreeable termination of the downstream view of the river just before it rounded one more bend and vanished. And crucially for my state of wellbeing, for my immersion in that symphony of nature, that lesser fall contributed the perfect acoustic counterpoint to the song of its upstream kindred spirit.

All afternoon, a non-stop procession of mayflies flew by, sometimes in dense clusters thousands strong, sometimes strung out in mere hundreds. Fly-fishermen the world over emulate them with flies of their own devising, but then I never did understand fly-fishing. The flies themselves caught fire as they passed me and I turned my head to follow them into the sunlight, and they were yellow and gold sparks and they thronged the first dozen feet of airspace above the surface of the water. From time to time one would land on the bright rectangle of my open notebook and bask there for several seconds, only moving on when the nib of my pen approached too close, and it transformed again from something grey and black into one more gold and yellow spark. If you have never thought there could be beauty in a fly, you need only watch a newly hatched horde of mayflies trekking upstream into the afternoon sunlight above the course of a Highland river with a song in its step. Only once in the whole afternoon was there the hard slap of a splash, but my eyes were on my pen and the notebook page at that moment, and when I looked up there were only widening ripples, so I never saw the rise, the bite, the fish-glitter, the supple twist in the air and the splashdown.

I left the river when the wind strengthened and cooled and I began to hunger and thirst, and I was healed again until the next time. But no wild Highland river has ever failed me yet.

Chapter Four

The Mountaineering Badger

IF I WERE TO follow the river for a few upstream miles, as I have done by car, bike and on foot surely a thousand times by now, I would come to the base of a mountainside that clasps a secret world to its mutilated body, nothing less than a new wilderness. I have lost count of how many times I have threaded its chaotic slopes, almost always in April or May, how many times I have celebrated the eccentric beauties of a Highland mountain spring that hardly anyone sees. For although this mountainside is the same age as all the others in the glen, sculpted and steepened by the same glacial hand, in an extraordinary way it has been born again. And there, 2,000 feet up the mountain, shining in the sun like a ragged mainsail, is its birthmark – its re-birthmark.

What happened was that about thirty years ago, and for no known reason, the mountain burst apart. From the transient perspective of a single human life, we tend to think of mountains as symbols of permanence, of immovable stillness, all the more so in a country like mine whose geological history of the last 10,000 years since the great ice relented is devoid of volcanic twitches, added to which is the comparatively new knowledge that the mountains of Scotland are hewn out of the very oldest rocks in Europe. The cliché we reach for to represent great age is "as old as the hills". Yet mountains move. They shift, they shrink, they rise, they respond in innumerable ways to nature's irresistible, invisible forces.

The buttress had been triangular-shaped and dull grey and from far below it looked...as old as the hills. I knew there were cracks in it, some of them wide enough for my hand, one or two where I could get an arm and a shoulder inside. I knew, too, that just below the buttress, on a wide and level shelf, was a boulderfield that was a consequence of some old upheaval in the mountain's story. But when I heard that something of the mountain had burst apart, I climbed it again to see what had happened, and what I saw first was that something shone in the sunlight like a sheet of tinfoil. It was a wound. A huge wedge had fallen from the buttress, and the boulderfield on the level shelf began to move. A new rockfall began. But lying in its path was the one thing that prevented a much more epic landslide – a mountain oakwood, an oakwood that had already adjusted to at least one landslide, thwarted its descent at the cost of many broken trees, but which, in the manner of woods left to their own devices the world over, simply settled again to the new circumstances and grew more oaks, and these hemmed the boulders in.

So now there is a steep mountainside, an oakwood that climbs from the floor of the glen to somewhere around 1,500 feet and in which there is barely a square yard of level ground. The underfoot conditions are so treacherous, so fankled by rocks and boulders and great slabs of mountain rock, that very few grazing animals, specifically deer and sheep, care to venture within, and so the wood and its rich and varied understorey thrive unmolested. Yet there are paths. They are narrow and steep and they curve crazily, and some disappear under fallen tree trunks, others where two rocks have come to rest against each other. Following any of them for any distance is one of the chanciest pursuits in the length and breadth of what I think of as my nature writer's home territory. For although the paths climb the mountain, they were not made by people.

Somehow or other, sometime or other (the pursuit of biological truth on this of all mountainsides is an imperfect and inexact science), badgers established a sett here, and as you might imagine, the inhabitants don't always do things the way field guides say that badgers do things. Here, for example, is where I have seen a badger front-pointing up a rock face as near to vertical as would make very little difference to you or me, but fissured and splintered just enough for a badger with built-in crampons. The confidence of the climbing suggested two things to me: one, that this was an activity with which that particular badger was intimately familiar; the other, that she understood perfectly the wisdom of the first principle of rock climbing for humans – at least three points of contact at all times.

I was amazed because I had never seen it done before. I knew there were preposterous claims out there in some of the murkier depths of badger folklore and I have read a few that would make a saint curse in exasperation. But I had never heard of rock-climbing badgers, so the first time I saw it was through a vague haze of ever-so-slightly surreal disbelief in what my eyes were telling me. I was further amazed by the ease with which the feat was accomplished.

Three times in my life I have been on good rock in the company of good rock climbers (I was never one of those and never aspired to be one) and what impressed me most each time was an apparently instinctive sense of perfect balance. I know it was not taught because I asked all three and none of them could account for it; when they started rock climbing, it was there already. If I could interview the badger, I imagine the answer would be the same.

There is also this: she didn't have to climb the rock face, she chose to. There was an easy alternative, a natural ramp that cut away at a gentle uphill angle from the ledge where she began to climb, and at the end of the ramp a clear narrow track angled back and up to the top of the rock face. I established over time that both the ramp and the track were worn smooth and devoid of vegetation by nothing more than the regular passage of badgers and the occasional fox, and I could walk effortlessly up the ramp and the track.

Spring was always my preferred badger-watching season, not least because of the arrival above ground of new cubs, mysterious little laugh-out-loud hovercraft-like creatures, with a miniature badger face at one end, that traverse their nursery terrain as if they are never quite in contact with the ground. And not least because of the badger's apparent love of bluebell woods. And not least because in my particular neck of the woods, badger-sett evenings and late nights were invariably sound-tracked by the roding riffs of woodcock flight as they patrolled their territories.

But the badgers of the broken mountain seemed to move to the beat of a different drum altogether, a lifestyle dictated by the landscape surroundings of the sett. Why they chose it at all used to baffle me, until one night it occurred to me that every time I went there – every time I still go there, for that matter – there was no human presence other than my own, and no sign of anyone having been near the place in my absence. Besides, my own presence is a thing of utmost discretion, silence and stillness (that and that alone explains why I have had a badger pee on my wellies while I was wearing them).

It was the third or fourth time I had watched this singular badger climb the rock face that I was struck by a peculiarity of her climbing technique. If you or I climb a rock face, we reach up for hand holds above our hands, sometimes to the very limits of our reach. A badger can't do that, can't reach far above its head, and this particular badger didn't even try. (It was always the same badger I saw rock climbing, easily recognisable by an irregularity in her left hind leg, perhaps an old wound.) With its head close to the rock it found holds with its front feet right in front of her face – holds I could not see, incidentally. I never once saw her try to raise a foot above eye level. So instead of pulling with hands and arms and pushing with feet and legs as you and I would do, she was effectively pushing with all four limbs, always small movements but always quick and utterly flawless, so that the effect was of a seamless ascent that reminded me effortlessly of the way a cub crosses level ground. She climbed without a pause, and while she seemed to place her front feet with great care, the hind feet found grip and thrust instinctively – the balance thing again.

I came across the phenomenon for the first time one warm May evening when I had climbed up beyond the wood, almost to the source of the landslide, and borne on a warm wind from below, the delicious hyacinth scent of the bluebell wood in full bloom painted the air with every shade of perfect spring that I can ever imagine. I sat longer than I intended and the woodcock had already done a few laps of his idea of his own portion of airspace, and each time he passed, his croaky, squeaky voice sounded like the first few notes of Ellington's *C Jam Blues* so that I got the piano riff stuck in my head like something you can't shake off after two minutes in the elevator of a posh hotel.

When I finally decided to descend, I picked a bouldery way down through the wood, for I knew by now there were badgers on the mountainside but I was short on details, like where. I had begun to suspect that the rock face with the ramp might have something to do with it, and headed for a distinctive larch that marked the point where a badger path crossed open ground between the rock-fall wood and the larch plantation (badgers in there too!). At the last moment, I decided to see if I could descend onto the top of the rock face and watch the ground in front of it from up there. I was within a few yards of the top edge when I heard unmistakable badger voices from below. I listened hard. Two badgers, and it sounded as if they were on the ramp, which could mean that any moment one of them, or both, might appear on the track up to the top of the rock face from the end of the ramp. I crawled the last yard and was about to peer over the edge when there was a yelp from below and the sound of badger feet scrabbling on the rock below. I was still pondering how that could possibly be when a badger face appeared from below the edge and stared at me from eight feet away. The animal dived downwards and sideways and disappeared into what I thought was unyielding rock, which gave me something else to ponder. I looked over, saw nothing, except that there was a second badger on the ramp, looking up. In an instant I saw its scut disappear, again apparently into unyielding rock.

I had caused enough mayhem. I left with as much haste as I could muster, making it clear that I was going.

The next day I was back, and with the wind blowing my scent away from the rock face, I took a long hard look at it in good light from a flattish rock with a helpful screen of birch and hazel. Because in a badger haunt as untypical as this one, you just never know. There was a hole at the bottom of the rock face, and I now realised that the rock face was actually just one side of a house-sized boulder. Instead of excavating a tunnel and a series of chambers and with a huge spoil heap for a doormat, the badgers had simply dug into the space beneath the boulder; there was no spoil to speak of, and the hole they had dug was narrow from top to bottom and wide from side to side. Nothing about it looked like a badger sett. But it was in there that the second badger of the night before had disappeared. And I now saw how the first badger had disappeared, although I still had trouble believing that it could have climbed the entire rock face from the sett entrance to the top edge. There was angle in the face, to the right of which there was a long crack several inches high, and the badger had made a scrambling descent into the crack. Later I would discover that it offered a kind of hidden fire escape onto the open hillside and the path to the larches.

From then on, I simply watched from the birch and hazel screen, and from there I saw badgers emerge from the solitary entrance (at least it was the only one I ever found), and almost invariably turn right onto the ramp; *almost* invariably because every now and then the slender sow with the damaged left leg turned left and into the wall, and began to climb straight up. And it was then I saw how she climbed. And it was then that I realised that when it comes to the inner workings of life in a Highland badger sett, I didn't know the half of it. Twenty years later, I still don't.

Part Two

Island Spring

Chapter Five

The Nature of Second Spring

I AM AN EAST-COAST mainlander from Dundee, a Lowland Scot thirled by birth to a sunrise-facing shore all but devoid of islands, and yet I thirst for the Hebrides and the Northern Isles as a desert wanderer thirsts for oases, and I hunger for mountains as a polar bear hungers for seal holes in Arctic ice. I used to wonder about the source of these cravings, occasionally marvelling at their power, until the day I discovered an ancestral tap-root in Donegal, whence my particular branch of the Crumley tribe emigrated to Dundee in the 1830s and sank new, fecund and enduring roots there. Then I stumbled across a fleeting reference to the name Crum as a sept of the MacDonalds of Benderloch, a fragment of the Argyll coast in thrall to the island of Lismore, of which its historian Robert Hay has written in *Lismore: The Great Garden* (Birlinn, 2009):

> *From the top of Cnoc Aingeal, the fire cairn of Lismore, you look northwards into the jaws of an Earth movement of unimaginable scale and age. Three blocks of crust, wandering over the surface of the planet, collided with such force that their edges crumpled upwards into mountains of Himalayan proportions.*

Suddenly, I recognised myself. Suddenly, I comprehended the forces at work within me that fashioned my priorities and inclinations as a writer about the natural world. Besides, in the seagoing heyday of the Lords of the Isles around a thousand years ago, Lismore to Donegal was no distance at all, and the back-and-forth between Argyll and the north coast of Ireland was run-of-the-mill.

So I began to feel more at ease with the realisation that there was a potent Celtic-Viking component in my DNA that tempered and occasionally shouted down the up-country Angus blend of Andersons and Barries with which my parents' mothers had also endowed me, not to mention a strain of Yorkshire Illingworths by way of my mother's father, and he was the storyteller among them all. It is true that mountains had always been on my horizon (on the right kind of day you can see Schiehallion from Dundee's landmark hill, the Law), but I have to acknowledge, too, that island mountains have long enjoyed a charm all their own in my heart of hearts. And finally I knew that the island addiction I have exhibited throughout my adult life, and which reaches from Muckle Flugga at northmost Unst in northmost Shetland to Orkney to Outer and Inner Hebrides to the Mull of Oa in the south-west of Islay (and down the east coast to the Isle of May and Lindisfarne), is well founded.

But quite why it took me more than forty years of island-going to make it to the Isle of Colonsay is a bit of a mystery to me. Even then, the occasion rather fell into my lap, for it came in the form of an invitation to the Colonsay Book Festival. For a Scottish nature writer impregnated with that island thirst, then, this was something of a dream gig. It also explains why on an April afternoon, watching the Hebridean world go by from a sun-laved deck of the Colonsay ferry, you might say that I was in my element. All around was the oceanic land of my ancestors, and as if by way of greeting the return of the prodigal while I leaned on the stern railings of the ferry, a sea eagle hauled itself up out of dark shadow into pale, lowering sunlight above the cliffs near Carsaig. And the sea eagle is also in the throes of re-establishing itself in the oceanic land of its ancestors, and this one rose into that island sky with such a heroic heft to the flight that I thought of it as an ambassador of what, in the specific context of this book, I might call the nature of second spring, that rebirth that is the evolving story of Scotland's reintroduced sea eagles.

◉ ◉ ◉

The memory of the first sea eagle I ever saw still haunts me after thirty years, and for all the wrong reasons. It was in a zoo. I despise zoos, all zoos, everywhere. I was visiting that particular zoo because I had been invited to attend a meeting there, the purpose of which escapes me now, although it must have seemed important at the time because I do remember protesting about the choice of venue and being overruled, yet I went anyway. I was barely through the gate when I saw the sea eagle, a huge grey curve on a perch in an enclosure, as morose a creature as I ever had the misfortune to see with my own eyes. Whatever the reason for the eagle's incarceration, its eyes were dulled by it and its spirit broken.

Happily, my intervening years are liberally strewn with sea eagles at large from the island west to the mainland east to the mountain forest north, and from Orkney to the Ochil Hills. But it is precisely because their raw presence across the whole unfettered scope of such land-and-sea-scapes and all their skies so reinforces the possibilities and the value of reintroducing lost species, and so invigorates a nature writer's imagination, that I have never relinquished the memory of that symbolic incarceration. At the time, I was simply inexpressibly disgusted. Over the years, that disgust has found its voice: that captive bird was denied *flight*, and that is to deny the bird its very eagle-ness. In exactly the same way, zoos in Scotland today incarcerate wolves as a visitor attraction and deny them *travel*. Long-distance earth-travel is the wolf's natural habitat, as long-distance flight is the sea eagle's. From time to time, usually when I am watching a sea eagle jousting with ocean winds or golden eagles, or simply devouring some archipelagic tract of sky with its giant stride, that brow-beaten zoo bird shuffles sideways into an eerie corner of my mind to effect silent rebuke of my species. And yours. For our species was wholly responsible not just for sentencing the zoo bird to life imprisonment, but also for wiping the entire sea eagle race off our portion of the Earth. And yes, nature conservation has brought it back, and it has begun to prosper again across many of its ancestral homelands; but still, vested interests within our species rage against it and demand that it should be physically prevented from behaving naturally – prevented from behaving like an eagle, its eagle-ness denied – whenever and wherever that behaviour happens to be inconvenient for our singularly unnatural relationship with the land. To be specific, sheep farmers and crofters want to be able to kill eagles that take lambs, whether living or dead.

Too many of us will never accept that attempting to manipulate nature is the ultimate folly of our species, the final, fatal human weakness. One sea eagle locked up in a zoo and silently haunting a corner of my mind for thirty years, is eloquently, unbearably symbolic of all that, and from time to time it still insists on its pound of flesh, even as the ferry cleaves its Colonsay-bound furrow through that all-but-flat-calm April afternoon.

The south coast of the Isle of Mull appeared in that light as a curtain wall of black cliffs, a souvenir of that geologically ancient range that was Himalayan in its reach. Precision becomes elusive as another billion years slip by and geology folds them away into the inner sanctums of the Earth and its oceans. Of the mountain summits whose ridgey silhouettes do still perforate the sky above the sea cliffs, only one of them is higher than 3,000 feet. Its name is Ben More, Beinn Mhor in the Gaelic language that named the map of this place. Some people who don't care about these things will tell you its meaning is on the dull side of prosaic – Big Hill; but others who do care will tell you it means The Great One, and I sail with them. Ben More dominates Mull's mountainous south in much the same way that 20,000-feet-high Denali dominates hundreds of square miles of Alaska, and the two names mean much the same thing. Scale supplies the context.

But viewed from the sea, these mountains of south Mull stand back from the ocean. The intervention of that dark wall of cliffs erases their sculptural shapes from the waist down. These disembodied mountaintops are curiously diminished from here, retiring, airy, spacious. But the cliffs, like all good curtain walls, are indomitable, a brotherhood hewn from the island's own raw materials, impervious to invaders, whether armies or oceans, and in this case accommodating to sea eagles.

A few minutes ago, a small, scruffy, bald man came and stood at the rail a few yards away and began to scan the cliffs, the inshore waters below them and the sky above them with binoculars, working systematically east to west, then back again. If his binoculars had been a motor car, they would have had a flying lady on the bonnet, an RR badge on the grille. I have what I consider to be good binoculars, and they are certainly the best I have ever owned, but I also have a slight knowledge of the market and my best guess is that his were from the outer edge of the optical stratosphere. I decided that was why he couldn't afford clothes.

A kind of squeal emerged involuntarily from his open mouth and he shifted excitedly on his feet but without moving anywhere. I looked where he was looking. I saw nothing. He squealed again, apparently obliviously as well as involuntarily. I tried to think where I had seen him before. Then it struck me: he could have been almost any character out of a Giles cartoon. I used to love them but hadn't seen one or even thought about one in forty years. But if Mr Giles had been standing where I stood right then, I think he would have been reaching for a pad and a pen and thanking the Colonsay ferry for dropping a sprinkle of gold-dust into his lap.

The Binoculared One was about five-feet-four-or-five. He wore a faded yellow shirt, both short-sleeved and too short of girth at the waist to accommodate a slightly pendulous stomach. As a result, shirt and trousers did not quite meet at the front. He might have been fifty-five; the little hair he had was very blond and stood on end, having been back-combed by the sea wind. His trousers, pale blue and prolifically stained, seemed to have pockets all the way down, some of them quite out of reach unless he sat down. On his feet were red trainers with white soles. It occurred to me then that while I was taken up with this rare sub-species of ferry fauna, he could well be watching a sea eagle, and given that he was watching it through binoculars in the same price range as a nice wee terraced house in Dundee with front and back gardens and a river view, I thought that I detected an internal twinge of envy at the quality of the image he was enjoying.

As a rule, I am not much of a student of my fellow man, and even less so in such a theatre of nature as the island-splashed sea south of Mull, but there was something likeably hobbity about this creature, so that I half expected him to turn towards me suddenly and exclaim:

"Gandalf, Gandalf! Good gracious me!"

But instead, when he did turn to me (there being no one else in sight on that portion of that deck at that moment), smiling a knowing, even confiding smile, what he actually exclaimed, in what I imagine to be an Oxford University professor-ish voice, was:

"Well, that deserves a cake and a cup of Calmac tea."

So that explains the tummy, I thought, then I wondered

if he thought Calmac was a species of tea like camomile or Lapsang Souchong, and if he did he was in for a bit of a let-down, for I've had Calmac tea and in all fairness it's nothing to write home about. As Gandalf would say:

"Tea! No thank you! A little red wine, I think for me."

And off he stomped, the hobbity one, to his cake and pot of Calmac, and doubtless to fill a pipe and blow smoke rings.

I turned back to the sea cliffs, pointed my suddenly inferior glasses a little higher than I had been looking before, and there I found her. She hefted high and low the most extensive acreage of wings anyone ever saw along that shore, wings like hung sheets of dark silk, wings like furrowed fields. With these she rose steeply from sea-cliff shadows into oceanic sunlight and at once she paled, for a sea eagle is dark only in dark shadow or in silhouette, and her upper wings and her head especially are often mottled light tan and pale grey that can look almost white in the right light. And then her tail shines like snowlight and her beak is the colour of daffodils, for all that it's as lethal as a bushman's machete.

You would never think her beautiful in the way that a golden eagle can be, but at that moment, she achieved a singularly prepossessing presence even in the company of such a daunting land-and-seascape as the south coast of Mull, as if a fragment of the cliff face had broken free, become animated and taken flight. And always, when I have watched sea eagles newly airborne after a long stillness, I experience the same misguided notion that they have too much wing-span at their disposal, an illusion that seems especially true of the more massive, more unwieldy females.

⊙ ⊙ ⊙

The first time I ever saw the eagle-that-came-home flying free in the land of its ancestors was a May morning. I was on a hill road in the north of Skye. It was one of those days of fast weather that I always find energising – wind, sun and rain in perpetual and miraculous rearrangements that scatter rainbows as vivid and fleeting as kingfishers among the hills and shoreline bays. So my car-bound journey abandoned its original and lesser purpose, then abandoned the car in favour of a spontaneous sampling of the intoxications distilled by a characteristic blend of May and Skye weather. There followed a masterclass in the inexact science of putting on overtrousers in a gale (visions of them being wrenched from my grasp, soaring among the ravens of the distant Quiraing, and eventually washing ashore in the quiet aftermath of dusk somewhere around Applecross). I knew even as I wrestled with them that, sheathed in waterproofs, I would be too warm, but the alternative was really no alternative at all. Besides, this was a day on which the notion of being comfortable was abandoned with the car.

I expected ravens. Days like this seem to energise them too. The more it blows, the better they like it, the more you are apt to see them apparently flying for the sake of it, demonstrably enjoying themselves, often in loose, formation-free, aerobatic squadrons of half a dozen at a time. Such a squadron spilled over the hill skyline and dived down to meet me as I climbed, dizzying down past me not fifty yards away, hell-for-leather through sunlight and shadow so that they shone and darkened as they dived, as if some pulsing internal light source was activated by every other wingbeat.

Suddenly, I was immediately above them and looking down on their broad wingspans, a viewpoint that enhanced the on-off trick and made of it a kind of sorcery. This was why I had abandoned the car.

Then it became clear that I was not the object of their attention, for even as they powered on down, spilling hundreds of feet of mobile air from their wings in a handful of seconds, I saw past them in the binoculars to where a strange gargoyle protruded from the flank of a buttress, its head sunlit, but the rest of it lost in deep shadow. I had simply never seen such a head. When it craned out over the abyss to look up side-headed at the black bird-rain that poured towards it, the sun glared off a huge slab of vivid yellow, a thing of such size and colour that it suggested the bird had been disturbed in the act of trying to swallow an improbably large wedge of cheese, so large that it completely obliterated its bill. The realisation that the improbably large wedge *was* the bill was the moment when I knew I was looking at the eagle-that-came-home – the white-tailed eagle, the sea eagle, the fancifully-named "eagle of the sunlit eye" of countless press releases about its reintroduction first on Rum then at sundry other sites on both the west and the east coasts of Scotland. Wherever that translation of the Gaelic came from, it was surely a poet's source rather than a naturalist's or anything that was ever in common use for that matter. Seton Gordon wrote in *Days of the Golden Eagle* that the old Gaelic for the white-tailed eagle was *Iolaire-buidhe* – yellow eagle, which, if you were to see it in gargoyle mode on such a morning as this one, makes much more sense: that sunlit bill is a thousand times more conspicuous than that sunlit eye.

The first raven arrived and dived to within a few inches above the eagle's head, which flinched. The others blurred past, reformed below, banked, climbed and dived down again, watched every yard of the way by one sunlit and one shadowed eye. The yellow billhook waved ominously and gleamed dully in the sunlight.

There were four attacks in all then the ravens tumbled away down the wind to play on the pulsating updraughts, having satisfied some high jink or other. Then the eagle flew, and my fascination for its too-much wingspan began. The unfurling of an eight-feet-wide wingspan from a standing start was like watching a parachute open. As the first strokes rose and fell and the whole ponderous creature began to move, it simply looked as if there was far too much material to do the job. If your idea of what an eagle looks like is formed exclusively by golden eagles, as mine was until that moment, then the white-tailed eagle cuts a less-than-dynamic first impression. This one wheeled at once into deep shadow then slowly spiralled up into the sunlight. In sunlight it shines at both ends – brilliant white tail, bright yellow bill – and oh, those wings are huge. Popular culture, and especially tabloid journalism, seems unwilling to let one pass without reaching for some unlikely metaphor or other to convey the impact of those wings…flying barn doors, flying wardrobes, etc. My notebook entry for that first encounter read: "wings the size and shape and colour of a newly-ploughed croft". It had

the merit of originality, at least. The Danish-born ornithologist Bertel Bruun wrote that "the white-tailed eagle
does not impress the observer with the majesty of the
golden eagle. The silhouette is slightly disharmonic, with
a short tail and rather long neck suspended on a pair of
wide, almost soft-looking, board-like wings…"

Hmm, yes, but then it gets into its stride among sun
shafts and rainbows, flying dead level despite the pulveris-
ing wind and shining yellow and white at each end, and
it becomes clear that this is a bird designed by nature to
make an impact. One consequence is that the sea eagle
is never left at peace for very long. Ravens and crows
torment it mob-handed. So do swifts and martins, kestrels and sparrowhawks, gulls and fulmars. In fact, fulmars
accounted for the failure of the very first attempt to reintroduce the bird in Scotland – on Fair Isle in the 1950s
– by drenching the bird in that hideous bile they throw
up at anything and everything that infringes their nesting
space. The eagle's feathers were effectively ruined and it
drowned at sea. And, of course, it only ever had to be reintroduced in the first place because it was ruthlessly hunted
into oblivion by the Victorians.

I watched my first white-tailed eagle climb through the
morning. It headed for the hill skyline, its wings working
harder now – and yes, looking "softer" than golden eagle
wings. I can say that with some certainty now, for the bird's
next tormentors were two golden eagles. I was watching
the white-tailed eagle's progress up the sky through the
glasses when the first golden eagle gatecrashed the image in
the glasses and vanished at once and at speed.

I lowered the glasses for the wider view, saw one golden eagle circling above the still-climbing white-tail and another circling below it. Before that moment I had never given much thought to the possibility that I would ever watch a creature on a Scottish hillside that would make a pair of golden eagles look small. Enter the creature that achieved just that effect.

The higher golden eagle folded its wings and fell and as it did so, its mate passed it going up, an almost vertical power-climb. The diving bird unsettled the white-tail so that it flipped onto one wingtip, and again there was the impression of "soft wings" at work, of too much wing to execute the required manoeuvre. At once, that impression was revealed to be utterly misplaced, for, still on one wingtip, the bird turned in such a tight circle that it was able to launch a salvo of its own at the golden eagle that was now below it.

The golden eagle responded by flipping on to its back and presenting its talons, an invitation the white-tail declined by pulling out of its dive and driving away across the hillside, a retreat (if retreat is what it was) cut short by the arrival from above of the second golden eagle.

For the next quarter of an hour or so, these three birds tumbled across the sky and the hillside, and all the time the sun was in and out of the clouds, the rain charged across the face of the hill in brilliant squalls and rainbows danced among them all. How long it might have gone on for is anyone's guess, for it seemed more ritualised combat than a serious attempt to spill blood, and the manoeuvres so extravagant there was surely an element of ecstasy in the performance. But it ended in a thunderstorm, in a sudden

welling of black cloud, in a torrent of rain that sent me
– and, I suspect, the eagles – in a headlong spring for the
cover of overhanging rocks.

When it ended, it had dragged the fight out of the wind,
the sun bore down, and the black cloud headed across the
island for the mainland, towing one more rainbow behind
it. There was no eagle in the sky, or at least not in my portion of it.

I headed slowly downhill, re-running the spectacle of it
all in my mind, trying to come to terms with what I had
seen, and trying to imagine what the golden eagles would
have made of the first arrival on the island of a bigger eagle.
I would hear in due course that here and there on the Skye
coast, the white-tailed eagle had reclaimed traditional nesting sites that had long since been taken over by golden
eagles. But after that first encounter, I was thinking that
surely the golden eagles' inheritance – its race-memory
– included an awareness of the white-tailed eagle, and a
knowledge of what it was and what threat it would pose,
if any. I have proved to my own complete satisfaction that
evidence of inherited rather than first-hand awareness is
widespread among the tribes of nature. In places where the
wolf has been reintroduced, for example, prey species such
as deer know instantly what it is and how to behave in its
presence, even if it has been absent for a hundred years. A
badger sett I know was recolonised after it had been cold for
thirty years, and the nearest occupied sett from which the
new colonists could have travelled was more than twenty
miles away. When I asked a seasoned badger-watching vet-
eran about this he had shrugged and said:

"Once a badger sett, always a badger sett."

The reintroduced beaver in Scotland exhibits behaviour derived from the need to make fast returns to open water from the natural predators of its mainland European ancestry – wolves and wolverines and bears. The returning white-tailed eagles are all descended from Norwegian stock, yet they immediately acquired the knowledge of their own traditions as a Scottish species even after an absence of a hundred years. Many of the sites where they now nest are sites where they nested more than a century ago and for centuries before that. Once an eyrie site, always an eyrie site. As with any animal made extinct, the landscapes where they once thrived don't stop being suitable just because the creature is no longer in the landscape. Put the animal back, or create the conditions for it to return, and it knows where to go and how to behave when it gets there.

My next white-tailed eagle was on Mingulay, a wandering youngster. The next was at Talisker Bay on the west coast of Skye, where I saw it lift an eider drake from the sea. It flew low over a group of a dozen or so eiders so that they all dived underwater. The first one to surface was still gasping for breath when the talons struck, and the eagle didn't even get its feet wet.

My next was on Mull where I saw something like an old rotting wooden post standing on a muddy shore. Only my long familiarity with that particular shore persuaded me to look twice, for I knew that there was no wooden post there. The binoculars showed it up for what it was, a white-tailed eagle standing erect, apparently staring at a crab and wondering what to do with it. Eventually it lost interest and walked a few yards across the shore, looking (I imagine) a bit like a hobbit. Strange how hobbits keep insinuating themselves into a chapter titled "The Nature of Second Spring", but after all, at the end of Bilbo's journey, "the desolation was now filled with birds and blossoms in spring" and "the prophecies of the old songs have turned out to be true".

When the hobbit-eagle eventually flew and vanished from the bay, I crossed the mud to the area where it had been walking in search of something I had never seen before (and never seen since, for that matter): eagle foot-prints. I found them, too, and they were as larger-than-life as every other aspect of the bird's physique, and among the commonplace tracks of oystercatcher and heron and gull and sandpiper and the rest, they were – what was your word, Mr Bruun? – disharmonic.

Since then I have seen white-tailed eagles often on Skye and Mull, roosting in a hilltop wood in Perthshire, and low over my native landscape of the Tay estuary, not far from Dundee following the success of a release site for reintro-duced birds in the singularly un-Hebridean, un-Highland landscape of North Fife, for historically the white-tails also ranged up and down the east coast of the land in places where a golden eagle wouldn't be seen dead.

There is a prosaic aspect, then, to the habits of the white-tailed eagle, one quite absent from those of the golden eagle. They are much less wary of people, much more tolerant of what a golden eagle would consider disturbance, and you wouldn't expect a golden eagle to leave its footprints in the shoreline mud. But if you make your first acquaintance of the white-tailed eagle somewhere like, say, Mugdrum in the Tay near Newburgh, or wandering a low-tide beach among the seals out at Tentsmuir Point, be wary of jumping to easy conclusions or making false assumptions.

This is a bird with a vast repertoire. It's at home among Norwegian fiords, Icelandic islands, and in Scotland there is no landscape which you can say for certain is beyond the reach of its giant shadow. And be in no doubt: it is more than willing to dance with the mountain gods, to dice with golden eagles, and to joust with the fastest of rainbows. And it's maybe best if you don't let one hear you say you think its wings are soft.

⊙ ⊙ ⊙

Somewhere between the Colonsay ferry and the moment at which I sat down to write a book titled *The Nature of Spring*, it occurred to me that this reinvigorated state of affairs resulting in the desolation being filled once again with birds and blossom in spring has taken exactly a hundred years. The very last sea eagle of the original Scottish population was shot in Shetland in 1918, eight years after its mate had been shot, eight years in which it had returned annually to the same eyrie cliff where it waited for another of its own tribe to appear, but by then there was no tribe left. Two failed attempts at trial reintroductions occurred in the 1950s and '60s, but it wasn't until 1975 that a properly resourced scheme administered by the old Nature Conservancy Council began on Rum, the tide finally began to turn in the birds' favour, and a second spring dawned for the white-tailed eagle. So that spring of 2018, that was so ill-starred for so many of nature's tribes, was also something of a landmark, something of a history-maker, something of a breakthrough in the vexing annals of humankind's relationship with nature. The celebrations may have been muted, but even muted celebrations are worth raising a glass to: a little red wine, I think for me.

Chapter Six

Forty-eight Hours on Colonsay

A PAIR OF GREAT northern divers hove to a hundred yards to starboard, at ease on the quiet sea until the wake of the ferry found them and they rode its bright crests and its shadowed troughs, so that when they rose to each new wave the sun lit them from stem to stern. Hebridean seas embrace no more exotic presence, nothing more blessed by wild finery, nothing more luxuriously bejewelled than great northern divers resplendent in breeding plumage. I live these days in the middle of the country where Highlands and Lowlands collide, and just about as far as you can get from the sea in Scotland, as far as you can get from the realm of migrating divers, but that island thirst to which I am prey ensures that we are far from passing strangers.

These Argyll waters are vital to the great northerns. At its peak, the wintering population here can reach a thousand birds, one fifth of the total European population. At this time of year, the numbers probably include north-making birds from France, Portugal and Spain, and they stop off here because there are so many others of their kind scattered across these waters. And as one who feels a constant magnetic pull towards the northern places of our planet, it does not diminish their standing in my eyes to know that journey's end for the breeding adults within this fluctuating population is Iceland, or Greenland, or just possibly Canada. I watched that pair dwindle into the evening distance in the wake of the Colonsay ferry, and I wondered what far shore awaited them at the end of their voyage. Then I turned away to watch Colonsay take shape and draw near.

⊙ ⊙ ⊙

Colonsay ran the perfect book festival. Its scale was intimate, it was particularly well organised, it occupied a single weekend, and the programme was structured so that all the events were on Saturday and Sunday afternoons, leaving mornings for island wandering and evenings for socialising. Add to that the glories of the island setting itself, and you can perhaps understand why I awoke early on my first Colonsay morning, saw the sunlight and heard a cuckoo's voice drifting in the open window, and thought: "Two-and-a-bit days of this are never going to be enough."

An hour before breakfast, I found a track leading uphill from the hotel to the smallest stone circle I have ever seen, the central standing stone perfectly upright and attended by a ring of assorted knee-to-waist-high acolytes, slab-sided, sharp-angled, and on this particular morning the setting for a display chase by a pair of wheatears. But the hour belonged to snipe. They called from all over the hillside, that funny little tick-tocking noise, which, if you didn't know what it was, might induce you into a mind-warp that envisaged a Dahl-esque factory for the manufacture of metronomes. The whole jerky, disordered snipe-chorale was being conducted by the head snipe from a podium on the very apex of the standing stone. The embellishment of a non-stop cuckoo was also the kind of quirk that Roald Dahl might have come up with if only his travels had chanced this way at this hour: *Cuckoo Clarence and the Metronome Factory.*

I spent the morning on Colonsay's west coast, beside a huge tidal pool enclosed by low rocks apart from a narrow inlet from the sea. The water here was completely colour-less at the shore, but it evolved seawards through parallel bands of whitish turquoise, greyish blue, sky blue and royal blue to a final flourish of a blue so profound it was almost navy. The rocks enclosing the pool on the north side were well equipped with broad ledges, where belly-up, whis-ker-stroking, chin-scratching, all-but-comatose grey seals prostrated themselves to their sun god. Very little in the nature of Hebridean islands is quite so given over to indo-lence as a grey seal on a sun-smitten rock ledge. For the moment they were silent, but the pool's soundtrack was well served by an underscore of breaking surf against the westmost rocks, and by a kind of call-and-response routine set up between a score of eiders and a small posse of oys-tercatchers, a symphony in black and white with off-beat day-glo orange accents.

The combination of eider drake croon and island sen-sibility invariably reconvenes the memory of eiders on far island shores – a Shetland shore, for example. I was staying for a night in the Queen's Hotel in Lerwick after the launch of what is now an old and out-of-print book, *Shetland – Land of the Ocean*, a collaboration with photographer-publisher Colin Baxter. That dates it to 1992, yet the memory is vivid and undiminished. The hotel is one of those old waterfront buildings that wades out into high-tide waters. My room was on the ground floor – the sea floor – and as it was warm, my window was open. I was awoken at about 6a.m. by the "*Ahhwoooh!*" of an eider drake. I went to the window and the crooner was swimming about ten feet away. I lay back on the bed and listened to him for the better part of an hour then got up for breakfast, very content with the hotel's early morning alarm system.

Back on Colonsay, something was niggling away at the essential harmonies of what I had now decided was a pibroch for eider and oystercatcher set on a rhythmic ground of breaking surf. It was a series of unscored interjections, weirdly rasping croaks that had begun far off and pianissimo but were closing in with an insistent crescendo. And where had I heard that voice before? Then, as I tried to make sense of it all, a fast, rhythmic flourish of wingbeats wheezed cheerfully overhead and the sun lit up the glossy black plumage and scarlet accessories of a pair of choughs.

Then I remembered: two years ago, they tumbled down a sea wind at Harlech on an Atlantic shore in Wales, giddy and jazzy and whooping that same harsh wheeze, so that I took them away home with me in my head as ambassadors of that place, a defining presence. Twenty years before all that, I was camping alone at the south end of Islay, which is when and where I met them for the first time, not knowing then that they were on a journey towards the abyss, a road to nowhere.

Choughs have us baffled. Their decline resists our best efforts to help them back from the brink. Ornithologists fall out about the solution, if there is one. If you suggest a number of breeding pairs anywhere between fifty and a hundred, someone else will tell you that's much too high or much too low. The Scottish population is confined to Islay, Colonsay, and Colonsay's near neighbour, Oronsay. But why "confined"? There is nothing stopping them from spreading out to former haunts on Jura or Mull or Gigha or Kintyre. Nothing, that is, except that the one thing choughs seem to need in order to prosper is other choughs, and because there are no other choughs in these places, they don't go there. So the core area dwindles, the number of choughs dwindles, the number of chicks surviving the first two years before they are old enough to breed dwindles, and inbreeding within the diminishing gene pool only increases their plight.

Choughs don't make headlines. Not like endangered red squirrels or wildcats or reintroduced beavers and sea eagles, for example. Most people have never heard of them, and most of those who have heard of them have never seen one. Extinction, certainly in Scotland, is a real possibility, edging towards probability unless nature can convince them to feel unconfined again.

I had to go to work in the afternoon, if you can call anything as agreeable as a book festival on Colonsay "work". I carried the *joie de vivre* of choughs in my head on the short drive back to the hotel, and then I remembered their alpine kin that I once met at 9,000 feet on the Schilthorn in the Swiss Alps. One way or another, the chough tribe seems to delight in living life on the edge, which is just one more reason why I am drawn to them. And you have to admire their taste in landscapes, which is yet another reason.

⊙ ⊙ ⊙

Sunday morning, up to the stones again, pausing to gasp at the unfolding morning, for a pencil-slim horizontal line of dazzling gold lay on the ocean directly beneath the Paps of Jura, a trinity of mountains that scrolls Jura's indelible signature across the land-and-seascape of these southern Hebrides for miles in every direction. A sudden rain shower materialised out of a sunlit nothing-at-all and darted darkly across the face of the mountains, inhaling the reflected gold light from the sun on the sea, so that it was lit from within as it travelled. And immediately it reminded me of the flight of choughs, an effortless association of ideas that struck me as astounding: the visible world was a thing of turbulent grace and transience and light, and yet underpinning it all was (and what was your phrase, Mr Hay?) "…the jaws of an Earth movement of unimaginable scale and age". Such is the sorcery of which Colonsay is capable, a sorcery in which Hebridean landscape and seascape and weather routinely conspire to charm susceptible mortals. I am one such mortal. I was duly charmed.

Later, sitting with a coffee in the sun outside the hotel, with the percussive voices of snipe and cuckoo drifting down from the hill and the cheerful "*Zinggggg*" of yellow-hammers on the wires, I began to tune in to the mobs of starlings in a cluster of trees. The starlings were up to their tricks, for they are among the finest mimics of other birds. I heard them contriving buzzard calls and curlew calls from within their incessant burbling chatter. From time to time a real buzzard and a real curlew gave voice in passing. You could see the starlings cock their heads towards the source of their raw material, treefuls of Rory Bremners sharpening up their act.

There was never going to be enough time, not this trip at least. I was told that I must not miss *the* beach, Kiloran Bay under Carnan Eoin, Colonsay's less-than-500-feet-high summit. I found the beach all but deserted, by which I mean deserted of people, but a roll call of waders drifted in and out of the shallows, among them a solitary whimbrel. I always think whimbrels look like curlews down on their luck. They are somehow not big enough, not sleek enough, the bill is not long enough and not curved enough, and when they open it to sing...ugh! But I'm always pleased to see one, because I have never seen one yet that wasn't in an absolutely memorably beautiful landscape. Just like choughs.

The afternoon book festival sessions in the village hall had a tea break built in at about three o'clock (this is a *very* civilised book festival). A few of us took our tea and cake outside and sat or stood around in the sunshine, for the day had grown blissfully warm. Then there was a voice in the field next door.

"*Krek, krek!*"

Corncrakes!

Two calling males thirty or forty yards from where we lounged about drinking tea and eating cake. This book festival has corncrakes! Word got around fast. The hall quickly emptied and a hundred people gathered round the scruffy little field, people who had never heard a corncrake in their lives, and most of whom wouldn't be able to pick one out in an identity parade, but there they were, caught up in the euphoria of the occasion and setting, and that most dis-agreeable-sounding of birds won dozens of new converts to nature's cause.

Most book festivals give their authors a little goodie bag, things to remind them of where they've been. Mine included Colonsay beer and a jar of Colonsay honey. The beer was accounted for the day I got home, sitting in the back garden. But the honey...the honey was something else. It is, firstly, extraordinarily good honey, and I am a sucker for extraordinarily good honey. I used it sparingly over the next couple of weeks on good scones and good bread, and it was as if there was something fundamentally of the island in it, something other than just the endeav-our of bees and beekeeper. Or perhaps it was just me. The Colonsay in my mind amounted to much, much more than the sum of its parts, and as the honey was the one tangible product of its landscape that I could summon, perhaps I imbued it with some kind of ambassadorial role. I had made some notes during my stay (the nature writer's habit) and a couple of rough sketches (very rough, I am no one's idea of an artist) up by the standing stone, one of them with a snipe standing on the top. I had the memories these jogged, and I had the honey, and suddenly four lines of Norman MacCaig snapped into focus:

They wound and they bless me
with strange gifts:
the salt of absence,
the honey of memory.

Chapter Seven

Yell – No Need of Dreams

"Is this the last boat?" I asked one of the Yell ferrymen. It appeared to my inexpert eye that the timetable as advertised had acquired an unfathomable flexibility during the day.

"No...no..." he replied. His face was unsmiling and becalmed, and his quiet voice left longer gaps than seemed strictly necessary between the two "nos" and between the second "no" and the rest of the sentence, which was: "there will be boats again all day tomorrow."

He held my eye and I held his, and his was as grey and tranquil as Yell Sound at slack water. But then an Arctic tern flickered past and we both looked at it because it was laughing, and that was humour enough for the three of us (the islander, the mainlander, the flying globetrotter) so that when I looked back at the ferryman, we were all laughing. I would remember him years later on Barra, when I pointed out a similar discrepancy between printed timetable and actual departure to a different ferryman.

"Ach, timetables," he said, and with the same unsmiling and becalmed expression, "they're not to be trusted."

Yell is the second furthest north island in the furthest north cluster of islands in the land, which is why it needs two ferries – one to connect with Mainland Shetland, and one to connect with Unst, which is the furthest north island, although in my personal anthology of islands, and island springs in particular, Yell is out of reach on its own. But just so that you are in no doubt whatever about where you are, and if you are the kind of person who likes maps, you will need Ordnance Survey sheets One and Two. I have always inclined towards the north of the world and (all other things being equal) northward journeys within my native land, and there is no more fulfilling north-making journey in my native land than one that ends at Yell. Not even one that ends at Unst. It may be that that state of affairs came about because of the particular circumstances under which I made that journey for the first time. Uniquely for me, it was a journey that began in Kensington Gardens, London W2.

April is often a poor thing in Shetland, grey-faced and hollow-cheeked, withered by winter and flayed by the kind of ocean winds most of us never experience. The winter that preceded that particular April had been all gale, one which scarcely paused to gather breath. Snow, when it bothered to fossilise the rain at all, never lingered. There would be a tinselly handful of hours one morning to astound the Shetlander's unaccustomed gaze, then the grey blind dropped again and the wind bludgeoned the islands with redoubled ferocity. Meanwhile, when April arrived in Kensington Gardens, London W2, it was wearing shirtsleeves and it was warm. I had a brief morning stroll there, consoling myself after a futile interview with a publisher, a literary project stillborn (my first but not my last). It felt like a mortal blow to my writing aspirations. I walked in sunlight and birdsong

and I was briefly seduced. I caught myself thinking traitorous thoughts about life in London if I really wanted to write
for a living. It was what my countryman J.M. Barrie did, and
it worked for him. Then this happened:

In the late morning I flew to Edinburgh. Through the
afternoon I drove flat-out for Aberdeen where the P&O
car park seemed to double as graveyard for dead container
trucks, while all around, and no less incongruous given
the bizarre rhythm to which the day danced, cattle floats
steamed and stank and bellowed. By the following dawn,
the Shetland ferry was steaming past Sumburgh Head, I was
well-slept and showered and swithering between starboard
for the sunrise over the sea or port for the effect of that
sunrise on kittiwake cliffs. By nine I was hot-breakfasting in
Lerwick, and by noon I was easing my car from one more
ferry onto an island suspended in an eerie state of spring-
less April torpor; an island that looked like a becalmed
whale, a humpback (the hump is a moorland swelling called
Ward of Otterswick, a sniff over 200 metres), and I half
expected it to roll over on its side and raise a pectoral fin in
slow greeting. Welcome to Yell.

A mile from the ferry, I stopped the car. I stepped out
then simply stopped moving. I inhaled the peat-smoke-
and-salt-air scent of the islandness and the northness of
where I stood, and I recognised at once what was missing
in London, what would always be missing in London, a
kinship with a sense of place. In that moment I was done
with traitorous thoughts. I *would* write for a living whatever
it took, and I would do it writing about the Yells of the
world. The North.

Yell is not beautiful. It lies low in the water and it is more or less shapeless. And after such a winter it looked dark grey and dark brown, the moor shades. "Moorit" is the Shetlander's word. It looks perhaps the way that much of Orkney did in past centuries before all that agricultural greening began. The only green on a Yell moor in April is a boggy ooze. Yet Yell is the Shetland I carry with me, the Shetland I pack when I leave, the Shetland I am impatient for when I return. The reason is elemental. Rock, water, wind, wild. Oh, and that voice. That voice that drifted downhill from behind my back was so eloquent, so pitch-perfectly attuned to my idea of North.

In a strange and roundabout way, it put Sibelius in my mind, as the North often does. The Sibelius of his Symphony No.6 is the North in spring in my mind, a vivid distillation not just of the symphonic form (music critics generally agree that there is nothing like it in the modern repertoire) but landscape and wildness made music. It's Yell. He introduced it to the world in three words. He said that while so many of his contemporaries were inventing cocktails of outlandish colours, his offering was "pure cold water". If your mind is up to the task of distilling an entire island or an idea of North into a single wild utterance (it is no trickier than distilling the symphonic form into pure cold water), then that voice I heard at that moment was pure cold water from nature's symphony.

I stood about 150 feet above that inter-island channel of the sea known as Yell Sound, a sea bespattered with islands and skerries and moving light, a sea enlivened by the wind's ragged furrowing, so that it convulsed gently in slow, grey-white shivers, silvering towards the north. That fits perfectly into this synthesis of ideas the voice had liberated into the landscapes of my mind – those places to which I make notional pilgrimages while my feet are too mainland-locked for too long and too far south for my comfort. My travels brighten as they go north. They silver in the north like Yell Sound.

Far to the north, beyond the silver, beyond the northmost point of the Shetland Mainland, a line lay across the sea from west to east, a single thick ripple, beyond which the ocean's assault on Yell's west coast is no longer hampered by the Mainland's bulwark. Nature has the northwest of Yell all to itself, for the main road north makes a tactical withdrawal from that shore of storm and crosses the island to the east coast and the ferry to Unst. Beyond that singular divide, the ocean is changed utterly; a different symphony, a different North prevails there.

The voice had grown more demanding of my attention. It had come no closer, but it had become more persistent, repeating its subtle variations on a single short theme again and again and again, and for minutes at a time. I consulted the map. I was looking for a small lochan up on the moor within easy flying distance of the shore. A burn with the mysterious name of North Burn (there is no South Burn and it slips into the sea a mile from the south coast) seemed to lead directly to one such. I stepped up onto the moor, into the embrace of the raw stuff of the island. A mile later the land flattened and opened and a quiet grey loch lay in the lap of a gentle little hill. And there was the owner of the voice, and there swimming in circles round her was the owner of the second voice.

I knew red-throated divers of old, most memorably on south Skye and Raasay (where once I fell asleep listening to them from my tent and woke up to them six hours later). I climbed easy slopes to the summit of the small hill and there I sat and grew properly still and all Yell swam towards me, and it was all lochan-freckled moorland hacked about by long voes that ferried the sea deep into the island heart and the island's boatmen throughout history to sheltered moorings. Red-throated divers are emblematic of Shetland, for this is their one breeding stronghold throughout Britain. These moors, these lochans (Yell alone must have more than a hundred), and proximity to the sea from anywhere and everywhere are the red-throat's perfect habitat. In Shetland, it is more or less limitless.

A strange stain of pale greyish-yellow sunlight slanted along the flank of the hill and tinged the surface of the lochan. The birds' constant circling of each other and their crying voices are so bonded with my idea of Shetland that the one has become red-throated Shetland and the other has become the Diver Islands and that voice is the anthem of both. That light caught the widening ripples generated by their circling dance, until the whole surface of the lochan was palely glittering circles, and the birds' slow rotations were as the hub of a water-wheel. In the same way, or at least in a comparable way as I tried to make sense of the landscape and that moment in it, Yell is a perfectly seated hub in the midst of the Shetland archipelago without which (it sometimes seemed to me) the whole tapestry of islands would unravel into chaos. From any point of high ground on Yell, you see islands in every direction, and as you turn through 360 degrees you see the ocean respond to tides and light and shadow and wind, and sea and islands are a dance that forever encircles and leans towards the still centre; Yell, the heart; and at the heart of the heart the bird with the blood-red throat and the blood-red eye. It is a match made in Valhalla.

Back at the car, back at the nowhere-on-the-map where I had simply run out of steam and stopped, I sat below the road among grass and heather and stared with disbelieving eyes at the view from there down to the shore and the estate agent's sign that stood incongruously by a fence. There was a small croft house, with a corrugated roof that amiable shade of rust that suits old, instinctively worked stone so well, and three acres of downhill-tumbling field that declined from marginal at the top to seaweed and rock at the foot: it could have been mine, said the sign, for £12,000. But that was back then, the economic reality of my nature-writing life was on the downhill side of hand-to-mouth, and I didn't have £120 to spare. But a man can dream:

No Need of Dreams
On low Lowland days I'd drowse
by a hearth of dream-kindling,
would wake in the evening of the North
where silken sea-scent prevails.

On slow Island days I'd light
a fire on a cold empty shore,
would fan a flame to thwart
the hypothermia of its soul.

There I'd raise up old walls,
set down new roots, and my smoke
would call and call all
the old flame-fanning spirits home.

A crop of rock, rush, wrack and hush
would be my estate, I'd sing songs
and tell sagas to the otters: the spirits
who wrote them would smile

and join in from window seats,
from the rotting schoolhouse,
their voices would echo and echo
from a blaze of hearths.

On high Island days I'd drowse
by a dreamless hearth,
for as the old walls warmed again
there'd be no need of dreams.

Chapter Eight

An Island Pilgrimage (1) – Mull and Iona

Almost at once we were in a heavy sea, and the wind blew
with such force that all idea of crossing to Fionnphort was
quickly abandoned, and, sailing before the gale, the course
was set for the sheltered creek near the little fishing village of
Kentra. Things looked serious…each person had perforce to
lie at the bottom of the boat so that the ferryman might see
the more easily to avoid the overfalls of the heaviest waves…
But with April better weather would come to Iona and to its
sound, and there would come an end to the gales from the west
and the south-west – gales which on one occasion prevented
the ferry-boat from crossing for a week or more, so that no
intercommunication was possible between the island and the
far side of the sound, though it is scarce a mile in width.

Seton Gordon,
The Land of the Hills and the Glens
(Cassell, 1920)

"A CUP OF TEA, please."
  "Are you staying, or are you going on the ferry?"
  "I'm going on the ferry."

She looked out to where the boat had just begun to nose into the pier from its offshore mooring, rocking slightly as it advanced. Her brow furrowed (despite herself, it seemed, for she had one of those clear and unlined Hebridean faces with a brow that looked as if it was not given to furrowing) and after a little consideration she said:

"I'll better give it to you in a paper cup."

What is it about ferries that make philosophers of folk? Or perhaps she had just been reading Seton Gordon.

I wrapped my hands around the cup and was briefly startled by the heat. It really was a paper cup. But what the Met Office like to call "the wind chill factor" (a pompous-sounding cliché instead of saying "it's going to be a damned sight colder than it looks") reduced the tea to tepid over a hundred yards, by which time it had stopped warming me, inside or out. I sipped as I walked, tea-drinking pilgrim. Standing at the rail of the upper deck, I sipped the cool dregs and remembered. It had been ten years.

I was staying up in the north of Mull for a few days. Ben More the day before had been in truculent mood. Gales and stinging, sleety snow had forced a retreat from near the summit. There was no view, and the view is the whole point of that particular mountain, and don't let any Munro-bagging fetishist tell you otherwise. There is nothing in all Scotland, all Britain, to compare with the view from Mull's Ben More. And it was May. I was alone, I decided to regroup the forces of endeavour behind a rock. Coffee. A dram (a small one, a taste, purely psychological), distilled just up the road. It is one of the miraculous properties of single malt whisky that it ensnares the essence of the landscape that gave it birth, and right then, right there, that whisky tasted like stinging, sleety snow had been distilled and bottled.

A window in the storm, a seductive blink of sun, I hung in there for a little longer. The window slammed shut. The sleety snow thickened and stung. Down then, cold and defeated. But in the evening, May in the woods at Salen was blissful, warm, saturated in birdsong. I decided Iona next day would be my respite. Go as a pilgrim, supplicating my God whose name I spell Nature. I have written before that I discarded churches years ago, found what I needed in Nature. But Iona Abbey was extraordinary, and if the appropriate word for the defining characteristic of the place is not "spirituality", then I don't know what else it could be.

Columba? Hardly. He had been dead six or seven hundred years when this abbey was built, and it has grown derelict and been restored since then. His had been a timber thing, modest as the coracle that ferried him from Ireland under the horizon. Yet perhaps it is his gift that endures, his granitic faith in peace. Wherever else in the world that most flawed of ideals has faltered, on Iona it never has in a thousand years. It never has because of all who came after. Pilgrims. Sipping sanctity and cooling tea.

From the ferry railing with an empty paper cup in my hand, then, I remembered crossing the waist of the island to white sands, white as the summit snows of yesterday's mountain. And I remembered The Blue. Our planet plays host to innumerable shades of blue, and I consider myself fortunate to have seen my share of them. But when I saw the Iona Blue that first time, I was numbed by colour. It was the particular blue that begins at your feet where the sea lies in a pool as still as ice over white sand. From there it flows seamlessly through deepening shades and deepening layers of ocean. It is blue beyond colour, blue as spiritual experience.

Now, ten years later, standing at the rail of the ferry, going back to Iona on a day somewhere between the last of winter and the first of one more spring, and with a now crumpled paper cup in my hand, it was the blue I remembered; and there was Rum, far past Mull and far past Coll, Rum off the starboard bow. I had forgotten about Rum. And how small Staffa is.

This time I walked past the abbey. I thought I might save it for last. I wanted to climb Dun I, the lowly summit of the lowly island whose old Gaelic name is just that single letter, I. It means island. I wanted to see white sands and to drink my fill of Iona Blue. And suddenly I realised that it was one year since my mother had died. She was a religious woman, she called her faith her "rock", and at least I could understand that analogy, and there is no worthier place to pay tribute to rock than Iona. It's Scotland's first rock. It is older than most other fragments of the planet that many of us will ever step on. It was nature's foundation stone. Having laid it, she stood on it and began to lay down the rest, and to throw up other volcanoes to play with for another billion years. So my tea-sipping pilgrimage suddenly found an edge of purpose. Mum liked Iona. She came here once on a bus tour. You could not pay me enough money to go on a bus tour, but after Dad died, Mum had eighteen more years and with those she travelled, as far as Florence twice, and sometimes she just took the bus to the Hebrides. Now, for no reason

I could name, I associated the occasion and the islandscape with the date. I am no keeper of anniversaries, especially the anniversaries of deaths. I remember lives best, and I need no dates to celebrate and be grateful for the lives of both my parents. Suddenly, however, I thought that perhaps a quiet moment in the abbey at the end of the day might be appropriate. But Columba had something else in mind.

I climbed Dun I, bullied uphill by a westerly that would have shivered Seton Gordon's spine. I crossed the summit and sat in the lee of a west-facing rock. I could see white sand. I could see Iona Blue. I could see down the broad spine of the island. I could see up the island to Rum. And beyond that, identifiable by its ragged glimmer of snow, the Skye Cuillin. I waved, because that is how I greet friends-in-landscape. I descended to the shore where the wind filled my eyes and my hair with white sand and the Iona Blue was a song beyond music. I turned. I climbed back over Dun I. By then the summit was a wall of wind that took my breath away and rammed gallons of its own icy breath down my throat. I sat again. I sat this time because standing was too difficult. And sitting in that rage of winds, sand-smitten and colour-drenched, I found the utter calm which permitted the remembrance. Nature soothes the soul. A friend wrote that to me once.

I walked back to the ferry without entering the abbey. I wondered how many other pilgrims had ever done that. If any. My pilgrimage was honoured without it. Columba, presiding spirit of pilgrims, keeper of souls, or whatever, conjured the most vivid remembrance at Iona's disposal, not within the confines of a building but in the un-confines of nature. The abbey may be the focal point of what Columba's legacy has become. But it is I, the Island, which is the cathedral of this pilgrim.

"So, did you get to Iona today then?"
"Yes."
"What was it like?"
"Windy. Freezing."

It was too soon after the event to explain, too soon to think. I have since found this, its source quite unknown to me:

Part of the inheritance of the Celt is the sense of longing and striving after the unattainable and incomprehensible on Earth…Forlorn, he has the sense of fighting a losing battle for all his soul holds dear, for the simple life of old, for the beauty of the world threatened with utilitarian desecration, for outlived ideals.

That part of me which is Celt suffers from that part of the inheritance. But perhaps it must also take the credit for the eye that yearns towards the Iona Blue, and for a kinship with summit winds. And if, as I suspect, that inheritance as it affects me stemmed from that seagoing corner of Argyll in thrall to Lismore, it also seemed to me that there lay the next stepping stone for an idea that hit me between the eyes on the summit of Dun I, for it was an idea of pilgrimage, a pilgrimage for nature and one of which Columba might have approved. But Lismore lies on the far side of Mull, and no crossing of Mull should pass without quiet consideration, for unless you are an embryonic saint with a coracle at your disposal and you have voyaged from Ireland, it was ever a part of the pilgrim's journey to and from Iona. And for me, more often than not, Mull has been the journey's be-all-and-end-all, and among many, many souvenirs to which I fondly cling is this one, which was not without its own savour of pilgrimage.

High above the north shore of Loch na Keal, a six-miles-long gouge the Atlantic has torn from Mull's waist, a small clenched fist of ash trees bows perpetually eastwards, shaped and shaved by salted ocean winds. They are perhaps fifteen feet tall and they will grow no taller, but the wonder is that they survive at all. Nearby is the ghost of a blackhouse, dark-stoned and low to the ground, suffocating in bracken, drowning in its own dereliction, the usual deaths for Hebridean houses of a certain age, the ineradicable spoor of the Clearances. The blunt, low-lying island of Eorsa is moored out in the middle of the loch. Its profile climbs in flat tiers to a flat summit, like so much of Mull's landscape, the incline gentle from the east, abrupt as a staircase in the west, ending in a perfect right-angle and a 100-foot cliff. Its shape and dark mass look so implausible, as if has just been towed there and left for the night, so that every time I pause here after the inevitable absence of years, I half expect to find that Eorsa has been towed away.

I nurture a small ambition as far as Eorsa is concerned. As yet it is unfulfilled, but there's still time. It is to hitch a lift on a small boat some sublime day of late May or early June, or a canoe would be good if the day was sufficiently sublime, the weather forecast sufficiently unimpeachable, and if I ever learn to handle a canoe. The purpose would be to spend a day and a night and a dawn there, watching Ben More respond to what Yeats called

The blue and the dim and the dark cloths
Of night and light and the half light

and in the manner of the pilgrim. For doesn't the small island's flat summit sit deferentially at the mountain's feet like a supplicant, an island within an island, a stance devised by the parent island for the considered contemplation of its own Everest, its own Chomolungma, its own Mother Goddess of the World?

I recall an old April: new snow enlivened the top 500 feet of Ben More, the water pale off-white at its feet but becoming blue and bluer up the coast so that it was a royal shade out by Treshnish. The wind was a north-westerly and all knife-edges, bombarding the Western Isles with fast squalls. But on the shore of Loch na Keal it was merely persistent. There were only two sounds. There was the twiggy jabber of the passage of the wind through the ash trees (black-budded and utterly bare) and there was the soft, penny-whistle monosyllable of golden plover. There were two of them on a heathery shelf, dark against the sunlit sea, calling to each other. The call of a golden plover holds a quality of threnody that tantalises a human ear. For something so high-pitched and thin, it penetrates inexplicably deep, reaches improbably far back. I know, I know, it does no such thing. It is simply a contact call between two birds, but it is a dull soul who stands on such a shore in the company of golden plovers and hears only a contact call between birds. The Gaels, who by and large are not dull souls, call the bird *feadag*, which is also the word for a flute. *Fead* is "whistle" (or a high-pitched wind noise) and *feadan* is the bagpipe chanter, and whether they named the chanter after the golden plover or the other way round would take the mind of a Gaelic scholar, which I am not. But a chanter in the hands of a great player is an elemental force and just as affecting on a human ear as the spring *piobaireachd* of the golden plover, whether you hear the call-note's enigmatic fade or the slow-winged aerial song (a golden smirr).

The sun was over the mountain's eastern shoulder. West of Eorsa, Loch na Keal crumpled like tinfoil where it dazzled. Then the wind stiffened and chilled. A glance to the north-west showed that Coll had vanished. A driven stormed thrashed down the sea, swallowed Treshnish and Staffa, then bore down on Ben More itself. The thing sizzled over Ulva, crossed the mouth of Loch na Keal, as if a curtain the height of the mountains had just been closed. But on my side of the loch not a drop of water fell, though the temperature dropped and the rasp of the wind tormenting the ash trees drowned out the *feadagan* (at the last gasp their silhouettes flattened, beaks to the storm).

The mountain blurred.

But the sun still bored into the mass of mountain and storm, and for minutes more, the softened shape of Ben More remained intact.

The storm piled in.

The mountain faded.

The little island vanished, swallowed whole.

But the sun still bored into the mass and the mountain edges faded and brightened in and out of the edges of sight, until eventually it vanished and for ten minutes it stayed vanished.

But the sun still bored into the mass, pale and dim as clouded moonlight, and never did the storm snuff out completely its tinfoil patch on the water.

Then the wind eased. I turned and saw the fish shape of Coll resume its place in the sea.

The curtain peeled back, the mountain stood forward to resume its stance and its familiar shape, but it was brighter than before where the sun lit its topcoat of new snow.

The storm moved away east, less of a force to be reckoned with now that it had taken on the mountain and lost. In its path lay Lismore, but Lismore is moored in more sheltered waters than these.

Chapter Nine

An Island Pilgrimage (2)
– Lismore to Islandshire

LISMORE: AN LIOS-MÒR in Gaelic, The Great Garden. It's
the first thing you learn about the place. I once told that
friend who said nature soothed the soul: "One thing you
must do before you die – go to Lismore at orchid time."
The explanation of Lismore's fertility and profusion is a
geological fluke: it's a long, thin slice of limestone. The final
act of glaciation (final in the sense that nothing comparable has happened in the last 10,000 years, but in the next
10,000 years, who knows?) contrived an ice cap on what
we now call Rannoch Moor, from which glaciers flowed in
every direction. As these melted and the sea rose, Lismore
almost drowned. Rannoch's glaciers hurtled Etive granite
westwards then south-west down the Great Glen, which
explains why, now that the sea has finally settled down
(final in the sense that nothing comparable has happened
in the last ten millennia, but given the imponderables of
global warming, who knows?), you find erratic boulders
marooned on top of Lismore's narrow plateau. A hiatus in
the sea-settling process that began something like 8,000
years ago and lasted something like 3,000 years, accounts
for Lismore's collection of raised beaches, caves, sea stacks,

peregrine-strafed cliffs. Lismore is both a geological show-piece and an idiot's guide to the godly art of making land-scape where ocean and mountains meet. But for me, it was also a search for a glimmer of hope, a quest for confirma-tion that unlike, say, the winter of 2016–17, the spring of 2018 would actually turn up, however fleetingly, so that in the best tradition of pilgrimage I could raise my hands to the heavens, shout my hallelujahs and proclaim that mine eyes hath seen the glory of the coming of the swifts, or some other symbol of confirmation. When I set out from Stirling for Lismore, it was already April 23rd, and if you could have cut open a wedge of the wind it would have had "winter" lettered all the way through.

The reasons for choosing Lismore were twofold. Firstly, we are old friends given to occasional hail-fellow-well-met reunions at long intervals, mostly during old springs, and it has shown itself reliable enough in its capacity to harbour spring earlier and for longer than most of Scotland. Secondly, it feeds nicely into the notion of pilgrimage, especially once the decision had been taken to revisit that old Iona spring, Iona being something of an original source for pilgrims in Scotland, what with Columba and all that flowed from his endeavours there. So the writing mind followed the spoor of an admittedly selective thread of pilgrimage's history, so that a pilgrimage within a pilgrimage might unfurl, and that led directly to Lismore. From there, it would travel only slightly less directly to another island of particularly fond acquaintance, and in the process it would bind the west coast with the east, Scotland with England (or at least with that part of it that once thrust its northern border as far north as the Forth), and Columba with Cuthbert, the dove with the eider – Lindisfarne, one of the five parishes that once comprised the old Northumbrian fiefdom known as Islandshire. The idea germinated boldly, buoyed up by the considerable consolation that even if not so much a whiff of spring graced such a pilgrimage, there was always the renewing of auld acquaintance with friends-in-landscape.

⊙ ⊙ ⊙

These are the most familiar of roads. The car follows them instinctively. All it needs to be told is where to pause, where to turn off, where to stop. These are the arteries of the nature-writing territory I have explored for thirty years. But I had never seen them look like this. I left Stirling in late April, crossed the Highland Edge at Callander, past Loch Lubnaig and the Balquhidder Glen road-end, over Glen Ogle, through Glen Dochart, past the Glen Orchy road-end, across the Black Mount and the edge of Rannoch Moor, then into Glen Etive under Bucahaille Etive Mor, where I stopped and stared. All that way, the entire landscape was only the palest shade of grey away from being utterly colourless. Buachaille Etive Mor, in my mind the handsomest of all mainland Argyll's mountains, was pallid and weary, wrapped in rags of cloud shaded from black to hodden grey, and looking as lethargic as an old stag that had just tholed one winter too many. The only vigour to be seen anywhere at all was in water – every mountain gully and every burn and river and waterfall seethed white, and every wind-and-rain-whipped loch and lochan was dowsed in the same forlorn mourning shades as the mountain. Near the layby where I had pulled off the road was a dead black-face ewe, the perfect metaphor for the landscape where it had died: one more heap of widely scattered grey and black, symbolic of the mood inflicted on the land by the overlong and predatory winter. It had its teeth and talons locked into the corpus of what should have been prime springtime, choking the very breath of life out of the land.

Glencoe, that most galvanising theatre of Highland geology, was likewise as moribund as an east-coast haar in January, a grey shrug. All along Loch Leven the mood held sway. Then I took the Oban road for the Appin coast, rounded a long bend where Loch Leven gave way to Loch Linnhe; the sea brightened as it broadened, and the Earth awoke. It was as simple and sudden as that. Along the Atlantic coast the force field of the Gulf Stream had taken winter by the scruff of the neck, smacked it around the head a few times and heaved it into the stinkpit of dead seasons. In its place were daffodils ablaze on the verges, fat-budded rhododendrons (a presence that twenty minutes back down the road seemed as unlikely as pixie dust) beginning to burst into violent red and purple (all colour looked violent at this point in the journey), clusters of larches and birches that greened as they clustered, and – ye Gods! – palm trees adorned cottage gardens.

One more layby, a path through a thicket, birdsong that included willow warbler (Africa meets Argyll, a recurring miracle), and there in the midst of a sea view were old friends from the Colonsay ferry – four great northern divers, and close enough and well-enough lit for binoculars to paint in the black head, beak and neck; the neck with a white incised tattoo of tree trunks, as if an Ansel Adams photograph had been miniaturised there. I know, it's an extravagant image, but being giddy on the sudden confirmation that spring is not yet wholly extinct is an excusable circumstance for extravagance. So, as well as the tree trunk tattoo, the binoculars also articulated the white breast with wavy zebra stripes, black back with white-on-black grid of crossword squares and a smothering of white spots as vivid as a starry night.

These waters – and all the way from Argyll to the Northern Isles – are sacred to the nomadic tribe of great northern divers as they drift north through the spring, before making the leap of faith that commits them to journey's end in Greenland. These April-into-May days have seen gatherings of rare spectacle, including 417 off Kintyre in 2001, and two years earlier, 781 at Scapa Flow in Orkney. When I read about that, I thought: "I hope it was a sunny day." Imagine that starry-night plumage, multiplied by 781, daylight constellations, a mobile Milky Way of birds adrift on Scapa Flow. What an antidote to the blockships, the barriers, the wrecks, the war graves, the gun emplacements, and all the other wretched souvenirs of World Wars I and II: suddenly a sea fog is split apart by the sun and there on the water is an ocean-going fleet of 781 great northern divers.

That's why nature always wins. Always.

The day's destination was Port Appin, with Lismore a ten-minute ferry ride away in the morning. There were two eiders on the pier. Eiders have a talismanic presence in my life. They have a happy habit of rescuing me on lonely shores. They make me smile. They nest in circumstances that make me realise how little I really know about the singularly specialised fieldcraft of becoming landscape oneself – something I preach to others and try to practise myself. But then I would stumble on an eider duck on her eggs, "stumble" because her nesting takes me by surprise, because it is so perfectly done, the surroundings so eider-brown and discreet, the stillness so utter; and I feel like bowing in deference, in flagrant admiration, and for the way she would stoically meet my gaze. The colour brown was never so deftly nor so beautifully utilised to make such a perfect blend of function and art.

As for the drake, I cannot do better than Gavin Maxwell, who thought its breeding plumage suggested "the full dress uniform of some unknown navy's admiral". Sometimes the descriptive power of Maxwell's writing is heightened to a pitch that reveals his painterly credentials (he was a portrait painter before he gave his creative life over to writing). In *Raven Seek Thy Brother* (Longmans, 1968), his admiral analogy was followed by this:

The first impression is of black and white, but at closer quarters the black-capped head that looked simply white from far off seems like the texture of white velvet and shows feathers of pale scintillating electric green on the rear half of the cheek and on the nape; the breast above the sharp dividing line from a black abdomen, is a pale gamboge, almost peach. From the white back the secondary wing feathers of the same colour sweep down in perfect scimitar curves over the black sides, adding immensely to the effect of a uniform designed for pomp and panache...

I have gathered around me a throng of souvenirs that take the form and the sound of eiders. I have found them everywhere from Iceland to Shetland and Orkney, from Harris to Lindisfarne and the Farne Islands, from Mull to the Isle of May, and a 20,000-strong winter armada in the Tay estuary. Yet perhaps the most durable of all those souvenirs was that single drake in Lerwick, where its extravagant woodwind yodel woke me as it drifted in through my open hotel window. All alarm clocks should sound like that.

And now there was a welcoming committee of two on the ferry pier at Port Appin, conferring their blessing on the 21st-century spring pilgrim's arrival. There was a time – the late 19th and early 20th centuries – when the eider population in Argyll waters blossomed to such an extent that it became known up and down the west coast as the "Colonsay duck". Why numbers increased from an admittedly low start is not clear, but inevitably, once it became a ubiquitous presence, the human population began to plunder its eggs and the sumptuous cushions of its own breast down with which it lines its nest. The female on the nest is the ultimate definition of a sitting duck, for it is so unwilling to leave its eggs even in the face of the most blatant of threats, that she can literally be lifted by human hands, her eggs and down cleared from the nest, and then she can be replaced where she sat, all without so much as a syllable of protest. One way and another, the bird's own natural vulnerability allied to our constant tampering with the naturalness of coastal waters mean that as the 21st century advances, each new spring reveals a sharp decrease in nesting eiders. From oil pollution to shooting (to protect mussel farms) to the myriad consequences of warming seas, our species is forever inventing new ways to trouble the duck once befriended by saints.

The first thing to be said of Lismore across the quiet sound that late April evening is that finally I found somewhere that looked like it was taking April seriously. It was green! And the crags of its north end were girt with trees: limestone and ash, how they love each other's company. In the hotel that evening, I reminded myself of Robert Hay's vivid evocation of the significance of where I was going in the morning. At the very beginning of Chapter One of *Lismore – The Great Garden* (Birlinn, 2009):

> *From the top of Cnoc Aingeal, the fire cairn of Lismore, you look northwards into the jaws of an Earth movement of unimaginable scale and age. Three blocks of crust, wandering over the surface of the planet, collided with such force that their edges crumpled upwards into mountains of Himalayan proportions. After many millions of years of erosion, only their roots survive as the modest mountains of the Scottish Highlands, Norway, Greenland and the Appalachians... Even now, 400– 500 million years later, aftershocks of these events are still being recorded as earthquakes in the Great Glen, the most recent in October 2008 near Glenfinnan... This crumpling, squeezing, twisting and heating somehow combined to lift up, into the middle of the Fault, a slice of ancient limestone that eventually would form Lismore...*

The ten-minute ferry crossing from Port Appin eased alongside the slipway under Lismore's blunt and craggy northern pow with a gentle rumble, like an old lion. True, it was an inauspicious arrival for an ambassador from the future harbouring a notion of pilgrimage to the tribal heartlands of the MacDonalds of Benderloch and their mysterious sept, the Crums, who surely stepped this way time without number during the reign of the Lords of the Isles. Alas, there was no delegation from the Lordship to receive my oath of fealty. But I swore it anyway, under my breath so as not to alarm any of the other passengers, all but one of whom seemed to have nothing more auspicious on their minds than a gentle walk along the road to the island café then returning to mainland Appin, perhaps on the next ferry but one.

I had two destinations in mind. One was a quiet north-west corner of the coast where I could sit and watch a small cluster of islands and skerries that had "here be otters" written all over them. The other was Cnoc Aingeal, the fire cairn of Lismore, for the view "into the jaws of an Earth movement of unimaginable scale and age". There I thought I might stare into the 500 million-years-old birth throes from which – in its own good time – my own love for a mountain land drew succour. That was the plan, at least, but as with Iona, there were ambitions and expectations and then there were "events, dear boy, events".

At that pivotal moment in my journey into spring, the one for which my mind's eye had contrived a host of mountain landmarks that would befit the jaws of an Earth movement, the spectacle was not all that it might have been. Ben Nevis had presided over my vision, because that's what Ben Nevis does, for all that it is among my least favourite mountains. I have never climbed it and don't plan to, because it attracts hordes, and the hordes bring noise and rubbish and that was never the kind of company I care for in the mountains. But the Glencoe mountains and the lovely Beinn Bheithir of fond memories gathered in the east, and Ardgour (which has nothing to prove in that company) fringed the western shore. But for the moment, all was sunk deep in a singular shroud of thick, clinging, towering grey cloud.

Then there was a wheatear, pert on a rock, spotlit by a glint of sun, set against a blueing wedge of sea. The wheatear was widely known in Britain as "whitearse" until the name began to offend Victorian sensibilities and an absurd clean-up campaign resulted in one of the most thoughtless bird names in the history of ornithology. It's a bird of the hills and the wild open places. Ears of wheat have nothing to do with it, and it has nothing to do with them. But when it flies, and especially if it crosses a square yard of sunlight, its arse is unmistakably brilliantly white. This particular white-arse was the first I had seen that backward spring, which made it four or five weeks late. It would also prove to be the first of a series of spring firsts whose paths I crossed that day. Lismore yielded migrant birdsong, notably willow warbler and wood warbler. There were violets, primroses, wood anemones and milkworts.

On a cliff-edge, a mat of salt-wind-washed hawthorn clung to improbable life by sprawling over several square yards instead of attempting a trunk and limbs and branches, for which aeons of evolution had prepared it. Instead, it was nowhere more than seven or eight inches high. I looked for a central "trunk" or perhaps a root, but so dense was the flattened mass that I could not find one.

A kestrel held up against a considerable wind. Then, when it whipped away downwind, the lower slopes of Ardgour blurred agreeably through the binoculars. Then, looking down on the bird from a cliff-top, it was the sea that flowed beneath it. Strange to see land-lover windhover haunt the tide. This, I rebuked myself silently, is what is missing from your life: the intrusion of sea into the lives of everything in nature, from whitearses and windhovers to hawthorns and mountain ranges, to the geology of islands that can throw a limestone fluke into the heart of a sea loch and make it look as if it was planned all along, and to the way trees grow round a crag. That day on that island, pausing often to hug the spirit of spring close to me with longing and a sense of intimate kinship, I heard the words of one of my favourite jazz ballads drift through my head and linger there: "Spring can really hang you up the most".

Among the ashes were two whitebeams, and I have admired whitebeams for many years – from a circular stand of ten of them I pass on my regular morning walk for coffee and newspapers, to a startling intrusion high on a slope of Glenorchy of all places. Here, they flourished a crop of exquisite grey-green, ready-to-burst buds, and rose darkly from cliff-face shadow to cliff-top sunlight, at which point they obeyed the law of the land that decrees that all cliff-top trees will bow to the prevailing wind so that they leaned away east. But whitebeams almost in leaf: now we were really getting somewhere.

A perch on a level shelf of raised beach provided lime-stone furniture for lunch, and an uninterrupted view of that ottery clutch of small islands and skerries. In such a situation, "lunch" can occupy a couple of idyllic hours, even otterless hours (for so they proved). There was first of all the situation itself, the perch on Lismore's northmost crags, the foreground of sea that continued to respond to the sun's sluggish emergence by changing and deepening colours in horizontal bands that darkened by degrees into distance, Ardgour's handsome mountain profile that also responded to the change in the weather by shrugging off cloud and baring hill shoulders. There were grey seals out among the skerries, more great northern divers on the water, and two kestrels came and went and soared above the crags or above the cliffs and stalled on the air there, and not for the first time in my life I envied a hovering bird its view of the world below its airy stance.

The rising tide finally eased the seals into the water as their couches (also limestone) flooded, and with that, as if the evicting influence of the high tide had climbed the cliffs to the raised beach left by that old Rannoch glacier, the galvanising company of spring announced the end of lunch. A rough path dipped towards the shore, where the evidence of otters was omnipresent – a litter of spraints stitched with fishbones, crab claws and broken shells, and ground-down grains of shell so densely packed they looked like rockpools.

Then, from the shore at Port Ramsay, just as sunlight began to enliven the visible world, a familiar shape emerged from a cursory scan of a wood across the water, an erect grey-brown slab that looked too big and too heavy for the comfort of the tree where it appeared to have been hung, like a sheet left out to dry. More careful consideration revealed that it was not hung but perched. Then it raised its head from its breast where it had been rearranging its feathers with an implement that looked like a cross between a sickle and a banana, and instantly became a sea eagle. And there it stood and there it stared and there it settled into prolonged stillness, looking as if it might spend the entire afternoon in that attitude. It was a telling example of one of the bird's character traits that distinguishes it from the golden eagle: it has no fear of humankind and its works, humankind's settlements and humankind's noise. This one was in full view of Port Ramsay's street of houses in the middle of the afternoon while the residents went about their business. The nearest house to the tree where the bird perched was about 200 yards. A collie barked. Two people worked on a boat. Three more chatted in a garden between house and shore, their voices carrying far over the quiet water. Two cars appeared from opposite directions at the one road junction, stopped there while the drivers conversed through open windows, engines running. The sea eagle feigned disinterest, but it is a safe bet that it took in everything.

Not one of the people I could see appeared to know that it was there; either that or its appearance in that tree was so familiar that it had become part of the furniture.

I mulled over the bird's changing fortunes. One hundred years ago, almost to the day, the last of its kind in Scotland, in all Britain, was shot, and it was not as if no one knew that it was the last of its kind. Its extinction had been achieved deliberately. Now this, where a small island community appeared to be quite indifferent to the presence of one of the most astonishing birds in the northern hemisphere, casually perched on its doorstep. Mull across the water has even turned it into a tourist symbol.

A flashback barged into my mind. Twenty years before, I was in a small open boat motoring out from Hoonah, Alaska, a one-horse town in the Tongass National Forest. These were strange, eerie waters, lagoonishly glassy, lapping almost silently against shores of rock and mud, above which spruce and hemlock forest reached improbably far up mountains with no names, and extended for miles and miles. It takes the sudden appearance of a humpback whale a hundred yards off the port bow to inform a dislocated stranger that these waters are outposts of the Pacific Ocean too. As that dark-green shore glided past, tree by giant tree, we kept passing the erect, blond-headed totems of that country that are nesting bald eagles; they were as regular as milestones. My host, a part-Tlingit hunter and fishing guide called Floyd Petersen, cut the engine for a while so that we could talk more comfortably, with just the quiet accompaniment of the idling ocean. Suddenly a shrill, giddy voice poured a stream of molten silver down out of the trees, so that it bounced up at me off the water and hit me squarely between the eyes.

"What the...?!"

"Oh, that's the eagle."

Oh, *that's* the eagle.

I found him with the binoculars, just as he was poised to let fly again. He threw his head back, opened his throat to the sky, and out poured the silver-tongued deluge again, and up it bounced again, and a chill rippled across my shoulder blades. I thought of the opening bars of a concerto for wilderness. That image of an American bald eagle, head back and skirling is the one I carry in my head forever. Sea eagle and bald eagle are close biological kin, sharing the white tail if not the white head and the musicality. And the bald eagle is also unfazed by the proximity of humanity's habitat. The first one I saw was flying across a hotel car park in downtown Juneau, Alaska's state capital. But what the two tribes have in common most obviously is to stand erect in a conifer tree, sometimes for an entire afternoon. So what I saw when I scoured the woodland across the bay from Port Ramsay was a shape I remembered from a three-week expedition to Alaska for the BBC's Natural History Unit, twenty years before.

The patience of the tribe of wild eagles is a phenomenon of nature. It takes first-hand experience of it to begin to believe in its scope. In an unbroken half-hour of watching, this Lismore sea eagle neither moved nor uttered a sound. I have plenty of cause to recognise the signs after more than thirty years of watching eagles. In particular, there was an encounter with a golden eagle 1,500 feet up in a Highland glen. It was perched on a skyline rock. I was watching it from the floor of the glen. It did not move in four hours. The thing that troubled me then, and still troubles me now years later, is that when I acknowledged my own limitations and pulled out (quite apart from the four hours, I was three miles from the car and there was only an hour of usable daylight left), I would never know if he stayed there all night. Now, with all that feeding into my mind, I considered the stationary sea eagle and decided to head inland and uphill.

⊙ ⊙ ⊙

Lismore's small parish church is unexceptional. It was also turned back-to-front when the Victorians modernised the 14th-century cathedral. "Cathedral" is a confusing term: this was not Durham or Kirkwall or Lincoln. This was modesty, writ small. A reconstruction in John Hay's book shows a single-storey building like two semi-detached cottages with a tower at one end. But for all its modesty, it provides a 21st-century link that reaches back to Christianity's very first trade mission into Scotland. What connects Iona with Lismore and beyond is the same idea that underpins this entire endeavour of mine to beat a path through all four of nature's seasons. It is the idea of pilgrimage, a journey as a means to an end that binds destination to a particular quest for understanding.

The single name that leads the pilgrim and the historian from one island to the other is Moluag. Relatively speaking, he is the unsung hero of pilgrimage's saga that binds Iona to Islandshire, Columba to Cuthbert, dove to eider. Bear in mind that "relatively speaking" should not obscure the fact that we are talking about saints here, so by definition there is a degree of fame attached to Moluag in that there is sainthood. He was on Iona with Columba, and there is a tradition that they failed to extend saintliness towards each other. The handed down stories suggest at the very least that the two were competitive in their missionary zeal, that there was an unseemly power struggle over Lismore, and that Moluag won it by fair means or the other kind. It seems that even saints sin, albeit mildly.

But once he got his feet under the table on Lismore, his

stewardship is unblemished. He chose the site of a long-established Pictish religious centre for his own abbey and education centre. It seems he was always in Columba's shadow as Lismore was in Iona's, and his abbey never rivalled Iona in grandeur or trappings. The present church is nearby, as is the Sanctuary Stone from who knows which of Lismore's religious eras, and quite possibly it pre-dates them all. It is also known as Clach na h-Ealamh, the Swan Stone. So perhaps the swan was to Moluag what the dove was to Columba and the eider to Cuthbert. Perhaps. Robert Hay notes that in the *Martyrology of Oengus* Moluag is given this accolade:

> *The pure, the bright, the pleasant*
> *The sun of Lismore*
> *That is Moluag of Lismore in Alba*

Legends appear to overwhelm any glimpses of truth when it comes to the 1,500-year-old reputations of saints, but, given my own fascination for the tribe of wild swans, I like the idea that asylum seekers and other fugitives were immune to prosecution for a year and a day if they could touch the Swan Stone.

◉ ◉ ◉

I took the high ground on the way back to the ferry, found the kestrels again. They were hunting over the summit above the ash tree crags, tilting the land and the sea and the whole island to their bidding at the whim of supremely articulate wings, or hanging from the spread fan of their tails, wings beating time to the song of the wind.

How does that feel? To hover? I know it's a means to an end for a hunting falcon, but I sometimes wonder if, occasionally, it hovers just to enjoy being there, high up and looking round. Sometimes it's as still as the North Star. Sometimes it's as fidgety as a sparrow. Sometimes it tires of its stance and stops moving everything – folds up its tail, stiffens its wings – and miraculously it moves forward doing nothing at all, and then it stands still again, finds the wind, and is either fidgety or still. James Macdonald Lockhart wrote in *Raptor* of that moment:

The bird is standing still, walking into the wind.

Perfect. But is it ever up there when it doesn't have to be, when it isn't hungry, when it doesn't have chicks to feed, when it isn't teaching fledglings the artistry of kestrel flight? I suppose what I mean is, does it ever come up here to admire the view, up here above Cnoc Aingeal, the fire cairn of Lismore? I thought that if I were a kestrel up here, I would want to do that, and then I would like to sail north into what John Hay called "the jaws of an Earth movement" and flirt with the wind where it is no longer an ocean wind or an island wind but a mountain wind on first-name terms with the mountains and their gods. Then wheel at the narrows of the loch and sail back to this very rock, drift down to a perfectly controlled landing then turn and look where I had just been, and admire it, recall the thrill of the flight, just ever-so-slightly out of breath.

It was not the best day for the view. The jaws were clamped shut by a great smother of cloud. It rained there, as it so often does whenever Ben Nevis and Glencoe are in the frame, but Ardgour and Mull had begun to stir themselves, to shake themselves and stretch like a waking fox. And the sea in among those ottery islands and skerries had deepened to the deepest, darkest blue of all the sapphire shades. The West in spring does this, transforms moment by moment, to form new shapes, new colours, new moods, new scenes and acts of wild theatre, and all of it achieved with kaleidoscopic sleights of hand and flights of fancy, as if a kestrel were tapping the kaleidoscope.

A strange bird call on the air.

I stopped whatever it was I was doing and gave everything over to listening. The sound came from behind and below. It was constant, and there was more than one voice at work – so, the contact calls of a flock on the move? It sounded like nothing I could put a name to. The sea was grey down there, down where the calls seemed to be coming from, and it was only after a speculative search with the binoculars that I found a skein of seven greyish birds flecked with brown and white and flying in a perfect vee. They flew at wave-top level and they might have been curlews except that they did not sound like curlews, did not fly like curlews, and were too small for curlews. I thought about plovers but that didn't work either. They were a hundred feet below and 200 yards offshore, and they swept round the north end of the island and vanished. I was still scratching my head when the answer raised a cautious head over the steep cliff edge. They had landed just beyond the point where they had disappeared. The head was that of a compact, striped curlew-ish bird, and the beak was that of a compact, straighter, blunter curlew beak. It was a whimbrel, last seen on Colonsay two years ago; and two years before that I had been overtaken by two as I drove north through Caithness en route to Orkney. A flock of seven was the most whimbrels I had ever seen at once, a rare event and getting rarer. Its Shetland stronghold is in dire decline, and if they cannot sustain a population there, Scotland may lose them altogether. Just like the choughs, like the wildcats.

A single hawthorn tree stood on the cliff edge. It was about five feet tall, but its limbs were so writhing and wizened that it might easily have been fifty years old. Its trunk was no more than six inches high, whence it despatched four skinny limbs more or less upwards, and the whole thing and its sparse foliage bore the shape of the wind. It was a cruel twist of fate that planted its seed there, on the very crown of the very rim of the cliff-top. It stood out against sea and sky, and hardly against the land at all. Its stance was as precarious as the tree is tenacious. Within its modest dimensions, it seemed to be flourishing, a mote of life in a vast land-and-sea-and-skyscape. I wished it a long life. I have thought of it many times since then, and I have envied its place on the map of the world. Lismore has always treated me well. This time, it treated me better than ever. I could understand, perhaps, how earthly competition had briefly overcome the saintlier instincts of Cuthbert and Moluag when they first saw the possibilities of the Great Garden.

☉ ☉ ☉

One of the earlier pilgrims to Iona after the death of Columba in 593 was a deposed king of Northumbria called Oswald, although whether he cloaked himself in the guise of pilgrimage rather than what he really was (a king on the run looking for somewhere to hide) is unclear so long after the event. But during his stay he was impressed enough by what he found there to nourish the hope that he might convert his own kingdom one day. That day arose in 634 when he was given his crown back. One way or another, that year was a turning point in the pilgrimage business. At Iona he had been impressed by a young monk called Aidan, and it was surely no coincidence that in that very year, Aidan founded an abbey on Lindisfarne. By then Aidan had already founded Melrose Abbey, so now the eastwards migration of Christianity had crossed the entire breadth of the land, and Columba would surely have approved of its final destination on Lindisfarne.

Meanwhile, in Dunbar (now in East Lothian but then a northern outpost of the Kingdom of Northumbria) a boy was born – also in 634 – in whom all of the above would coalesce, and who, all but 1,400 years later, would capture my imagination. His name was Cuthbert. One night when he was seventeen (and here it may be advisable to have some pinches of salt to hand) he saw a light fall to earth, pause there, then retrace its steps heavenwards. It was, he assumed, lighting the way for a human soul to ascend its way into heaven. It transpired subsequently that Aidan had died that night. Cuthbert did what any susceptible seventeen-year-old mortal who has heavenly visions would do – he went to Melrose. He became a monk. He was good at it. In time he became a prior, and the next step would have been abbot, were it not for the fact that at thirty, he moved. To Lindisfarne. For the next ten years his reputation grew. But then, he either got fed up of the celebrity business or he felt called to become a hermit. One way or the other, he became one, and established his hermitage on Inner Farne, and from then on, he got on extraordinarily well with eiders, which is one of two reasons why I rather liked him.

The second reason is a little more complicated than that, and it may take a while. It will help if I reintroduce you to the writing of Seton Gordon. Why Seton Gordon? Again? Because he is the root, the original source, of all modern Scottish nature writing, the founder of the tradition in which I have laboured for thirty years now. And writing in *Afoot in Wild Places* (Cassell, 1937) he offered this, which explains in a moment what truly linked Iona to Lindisfarne.

South of Berwick-on-Tweed is Islandshire – that part of Northumbria in which lie Holy Island or Lindisfarne and the Farne Islands group. Islandshire is, even at the present day, a lonely district, and it is perhaps the only county of the east coast where is to be found what I may call the Hebridean atmosphere.

There it is. For the heirs to Columba of Iona, Lindisfarne was a home from home. I turned up there one spring day as a teenager on a bike. Periodically, and at long and irregular intervals, I have been doing it ever since. And when spring became the theme of my next book, like the cuckoo on the Mediterranean olive tree on page one, I looked at the calendar and thought: "It maun be time." It was a strange thought because south-making does not come naturally to me. It never has. My roots are planted on the north shore of the Tay estuary and I have never made any secret of the fact that my landscape preferences lie between the compass points of north and west. Stir into the mix my innate Scottishness (I was born here and I have lived and worked here all my life), my trade as a nature writer, and the knowledge that if I travel south nature will be in better heart where I have been than where I am going and all my preferred landscapes will lie behind me rather than ahead of me…consider all that and you may understand that any crossing of the English border is an undertaking that teeters precariously on the outer edges of equanimity.

And be in no doubt: Scotland into England is not one region of Great Britain into another. That was never how I interpreted the lie of the land. Rather, this is international business, a border between nations. This is France into Spain, Austria into Italy, Denmark into Germany, Canada into the USA. It is not that I am blind to England's many natural charms, rather that I bring to bear on my appreciation of them something of Lord Byron's comparative assessment set against his childhood memories of Lochnagar:

> *Though Nature of verdure and flowers bereft you*
> *Yet still art thou dearer than Albion's plain,*
>
> *England! Thy beauties are tame and domestic*
> *To one who has roved on the mountains afar –*
>
> *Oh! for the crags that are wild and majestic,*
> *The steep frowning glories of Dark Loch na Garr.*

Such was the old familiar preoccupation of my state of mind as another mile of the A1 slipped by, another bend in the road was followed by a crest in the road, and then there it was, without warning or fanfare, ten miles away and thrusting palely out into the grey-green sprawl of the North Sea, there lay Lindisfarne. And after more than a hundred miles of grey and featureless sky, the island stirred in a halo of weak but brightening sunlight. And right then, right there, as the English border hurtled towards me at sixty miles per hour, I felt love and longing, for I have been there before and loved what I found there.

Besides, I trust Seton Gordon. I always have. So when he writes: "...the only county of the east coast where is to be found what I may call the Hebridean atmosphere..." I believe him. His assessment matters to me, partly because he knew more about the Hebridean atmosphere than most people and articulated it thoughtfully and sometimes unforgettably in many of his books, not least in *The Charm of Skye* (Cassell, 1929), which is my favourite of his books, and was his favourite too. Skye lured me into what is now a far-flung appreciation of the Hebrides that has endured for almost fifty years. And for more than thirty of these, as a writer, I have travelled a line of latitude coast-to-coast across Scotland from the Tay estuary to Mull and it has been a fertile seam for me. Among much else, it has produced this:

Mull Ferry
(Arrival)
Waves in fractured parallels
stacked seven shades of grey
among old rock familiarities
that jigsaw into place; See!

That piece is me,
fitting back into its empty space
in the geology of my islandness.
Not that this is my native shore,

and I'm no prodigal here,
yet whenever I return
nature kills me fatted calves
and broaches flagons.

(Departure)
My life's other shore,
the native mainland one I ply
far from island ferries
insists on this retreat. See!

There's my piece again,
fitting back into the empty space
it left behind to indulge
my temporary islandness,

and where I am frequently prodigal
(while behind my back the island
wishes me safe journey,
haste ye back). And so I return

to that native place in my life's puzzle
but it's a light anchor I drop.
I stare out at a grey sea
without islands.

Somehow, in past readings of Seton Gordon, I missed that mention of "Islandshire". It's strange that I was unaware of its old name, because one way or another, Lindisfarne and its Northumbrian hinterland had quietly got under my skin, and it's something I would like to have known about the place. It is immediately obvious that it suits the place well, and it has endearing overtones. So I greeted the Border with a wave, because had I not just crossed out of Berwickshire into Islandshire? And didn't that remove some of the unhelpful echoes of old animosities from the Scotland–England thing?

So Islandshire (as I shall address it for evermore) has recommended itself to me for fifty years, and Seton Gordon has been among its ambassadors, and several of his books have been a constant presence through all my adult years and others have ebbed and flowed through. And Seton Gordon, in turn, came recommended to me first of all by an old uncle whose love of wild Scotland, birds, books, photography and cycling were effortlessly communicated to me and found a willing disciple. His name was Stuart Illingworth, he was my mother's brother, and he gave me my first bird book and my first malt whisky, so he was a considerable and benevolent influence on my young life. But he also introduced me to Lindisfarne when I was that teenager with a bike, roped me in with two of his chums on a cycling trek from his Peebles home to Northumberland, without ever telling me that part of it – the parishes of Ancroft, Belford, Elwick, Holy Island, and Kyloe and Tweedmouth – was historically known as Islandshire. I now consider this to be remiss of him.

He did, however, fuel my burgeoning interest in Scottish independence by talking mischievously about how an independent Scotland could annex it, bring it back where (as he saw it) it still belonged. And he had been a captain in the Lovat Scouts in Burma during World War II, and seemed (as I saw it) to have a grasp on the essentials of how it might be done. When I suggested the natives might put up some resistance (given that when King Oswald was in charge there was a time when Northumbrians overran the southeast of Scotland as far north as Fife, and that they might see the natural order of things differently), he swotted away my concerns as schoolbook history. But when I saw the place, when I considered the possibilities of not having to cross the border to get there – a Hebridean atmosphere in our far south-east – I confess I rather liked the idea.

Today, I still cherish two memories of that first encounter with Lindisfarne. The first was – of course – the causeway, for I had never seen anything quite like it. It was the speed with which the sea fell back, the road consolidating and shedding water like a surfacing submarine, before finally defining itself in a way that recalled an old movie in which Charlton Heston as Moses parted the Red Sea, the people surging towards the Promised Land along God's suddenly revealed causeway. Thinking about it now, what a way to end a pilgrimage.

My second souvenir was the first sight of the seagoing cluster of the Farne Islands.

Because we had come on it all from the west (not the north, as has been my habit in recent years, driving down the A1 rather than cycling from Peebles), and from high above, I saw how the land suddenly opened up and tilted all Islandshire towards that pod of islands; and hadn't there been something in schoolbook history about St Cuthbert?

Now the connections come thick and fast. A few years later, as I expanded my mountaineering horizons, I had become profoundly smitten by a passionate relationship with the Cairngorm Mountains, and my deepening awareness of the significance of Seton Gordon unearthed this in his book, *The Cairngorm Hills of Scotland* (Cassell, 1925):

In the immense silences of these wild corries and dark rocks, the spirit of the high and lonely places revealed herself, so that one felt the serene and benign influence that has from time to time caused men to leave the society of their fellows and live on some remote and surf-drenched isle – as St Cuthbert did on Farne – there to steep themselves in those spiritual influences that are hard to receive in the crowded hours of human life.

This interweaving of the 4,000-foot high mountain plateau and the sea-level island into a single philosophical strand took my breath away when I first read it. The idea that one writer could be so confident in his own knowledge of his native landscape and 1,500 years of its human history and make of them such a singular and striking image was new to me. When, in time, I wrote my own book about the Cairngorms, not only did I quote the passage (twice), and not only did I borrow from it (the book is called *A High and Lonely Place*), but when I elaborated on my own idea of what constitutes "the spirit of the high and lonely places", I reached for the example of St Cuthbert too:

The spirit is, I think, a collaboration of landscape forces, of light, of weather, of space, the mingled chemistry of which creates a tangible presence of nature that demands a response in those who encounter it. Respect for the spirit is the first commandment of the wilderness... The spirit lingers most blatantly on the plateau. There is no mountain sensation to compare with its sweeping plain, its corner-to-corner skies... the restlessness of its air that can wash great freedoms through a receptive human mind in the manner of St Cuthbert's surf...

So this land of Islandshire tilts towards the islands, and if you happen to come upon it from the west, your eyes reach down and out to them, and the chances are that you have just had your first glimpse of journey's end. For 1,500 years now, from Columba's Iona to Moluag's Lismore to Aidan's Melrose to Cuthbert's Lindisfarne (the "daughter monastery" of Iona), that way lies journey's end for pilgrims culled from every strand of humanity. For some of us, following in our own time and for our own motives (and perhaps especially the nature writers among us), there is arguably a sense of rightness about following in the spoor of a saint who befriended eider ducks; and given that in my own case others who had gone before also included my uncle and the man whose work he introduced me to – Seton Gordon – why would I not warm to the symbolism of it all and have it embrace the love-in-landscape that I cling to?

On a March day, then, that faltering backward spring of 2018, I arrived at the mainland end of the causeway to be greeted by a low-flying welcoming party of brent geese and their guttural benediction, which I chose to translate roughly as "hail fellow, well met". And this being the 21st century, there was also an impatient queue of half a dozen cars gnashing its teeth, drumming its fingers on its steering wheel and running its engine, as if collective irritation was a force to be reckoned with that might annihilate the gravitational pull of the moon and get the job done so that the sea would ebb away from the causeway that much quicker, and my carefully constructed illusion about the rightness of pilgrimage cracked and splintered and dematerialised before my eyes. What price a quest for those "spiritual influences that are hard to receive in the crowded hours of human life"?

A vast and bright red four-wheel-drive pick-up truck the size of a tug boat overtook the queue at speed, plunged into the retreating sea water, blasted a bow-wave that would not have disgraced the QEII in her pomp, bullied its way across. And the more timid drivers in that tetchy queue looked longingly at its dwindling tailgate and muttered in unison:

"I'm gonna get me one of those."

The brent geese, which have more sense (as well as wings, admittedly), flew over again no more than a dozen feet above the queue's throbbing roofscape, for it is nothing more than a fluctuating part of their landscape, that landscape which is always fluctuating in any case, always restless. Lindisfarne constantly tampers with its sea, and the sea constantly tampers with Lindisfarne, rearranging its dunes, and cutting it off from the rest of the world twice a day, then reconnecting it and baring tracts of puddled sand in the process, to which wildfowl and waders and seabirds flock in their many thousands and from all across the northern hemisphere. The brent geese, which are among the most restless of Lindisfarne's tribes, forby the fidgeters on four wheels, drifted away a little further from the queue with its fumes and its engine rumble, landed again a hundred yards away, causing a certain amount of readjustment among curlews, shelduck and godwits. I get impatient with impatient queues and idling engines so I retreated half a mile up the road to while away the wait in a café. I saw that it was almost noon. What was it about noon and godwits? Then I remembered. It was Seton Gordon:

> *The godwits decided at last that the wide estuary was unsafe for them, and rising to a great height they made for the open sea. Delicately pencilled against the blue of the noonday sky they constantly changed their position. Now flying in a close bunch, now in V-shape formation, they sped towards the friendly sea, their narrow, slender wings, fast-driven, recalling the flight of the peregrine. For a moment the flock turned as one bird and hung poised on the breeze. A second flock joined them, and the great multitude of birds which had temporarily become confused, changed into an orderly gathering that wheeled, glided and dipped as though directed by a single mind…*

It was from *Afoot in Wild Places*. There is a page marker in my copy, a small feather, and beside the passage in faint pencil and in my own handwriting it says: "Masterly."

So I sat with good coffee and a quiet contentment that spoke of journey's end. The sea wore a calm and muted sheen. But again and again and far offshore, a single wave would rise out of apparent calm, travel at speed over perhaps 200 yards, then break with a kind of controlled grace.

Moments later the next wave would rise up from nothing and follow it in. It became mesmerising. The small huddle of the Farne Islands crouched before them.

It was time. Lindisfarne, the Holy Island, and its Hebridean atmosphere awaited. As I eased my car onto the causeway a pair of eiders made way.

Chapter Ten

An Island Pilgrimage (3)
– Lindisfarne, Nature's Island

On a particular day in spring the weather will be perfect. The shining tidal pools of the Slakes, which bear the Pilgrims' Way, will mirror blue skies and scudding clouds. But even if the day should chance to be grey and wet, soaking the hundreds of folk descending the hill from Beal towards the Pilgrims' Way, in spirit they will remain undaunted. For they are pilgrims and the day is Good Friday. They walk barefoot, some share the burden of huge wooden crosses on their shoulders. They sing that their faith will remain constant, "come wind, come weather".

They congregate from all over England and from Scotland. They represent several denominations and have come to Lindisfarne, one of the holiest places in Britain, to re-commit their faith in this spectacular annual act of worship. They are joined by a few others, of no particular persuasion, but who sense, perhaps, that something missing from their lives might be retrieved, like a jewel in the sand, by walking the Pilgrims' Way to the edge of Holy Island at Easter.

Sheila Mackay,
Lindisfarne Landscapes
(Saint Andrew Press, 1996)

I HAVE LOOKED FROM both ends, and at both low and high tide, along the line of tall poles that marks the route of the Pilgrims' Way across the low-tide sands between Lindisfarne and the English mainland. It is a more or less reliable way to effect a crossing on foot more or less dryshod and – crucially – avoiding quicksands. I have to admit at this point that I am completely unmoved by it, unpersuaded by the need for the ritual, and for three reasons. Firstly, I am not *that* kind of pilgrim; secondly, I don't subscribe to that kind of faith (far less the urge to carry a huge wooden cross to prove…what exactly?); and thirdly, while it is unarguable that I can think of something missing from my life – that "jewel in the sand" I may or may not be able to retrieve hereabouts – it certainly wouldn't be retrieved on a public holiday in the company of hundreds of others. It is nature I seek an audience with, so if I were not to go alone, I would need the company of one singularly attuned to the notion that what is truly sacred is the land itself, and that nature is best met where the vast spaces of sea and sky conspire and collide with the edge of the island in the silence of low tide or the sound and fury of the cry of birds and the beat of breakers. I've written before of how much I like edges, and whatever else, Lindisfarne is an edgy kind of place, and its shores, especially its northern and eastern shores, constitute the edge of an edge.

Here is a tract of the shore where a half-beaten path the width of one pair of boots threads an uncertain trail through marram grass round the base of the dunes, and at low tide the sheer sprawl of dark sand looks as if the sea has been replaced by a flowing tide of tea-coloured beauty, the illusion of liquid assisted by wide tidal pools that declined temporary abdication when the rest of the sea took leave of its senses and vanished an hour ago. They have no colour at all, but they gleam, and crowds of crabs and shoals of other shell-thirled creatures flock to them or perish in the attempt.

A close inspection of the shore revealed tideline caches of feathers (eider, for sure, possibly short-eared owl), a handful of which now decorate my bookshelves so that every now and then I pick one up, run a daydreaming finger along one edge, and remember.

The day.

The hour.

The moment.

The atmosphere.

Or a draught from the window or the door lifts one from its temporary berth in front of the George Mackay Browns (I thought he of all the writers gathered there might enjoy their islandness) and drifts in an unhurried spiral towards the floor, just possibly brushing a guitar string on the way down, and I stop what I'm doing while I watch and let my thoughts rummage around the idea of a second spring for the feathers, so that however briefly, they fly again and delight the eye of the watcher.

Atmosphere.

That word again, in Seton Gordon's thoughtful obser-vation of eighty years ago: "…the only county of the east coast where is to be found what I may call the Hebridean atmosphere…". And if your particular pilgrimage can accommodate the time to tune in to such an elusive con-sideration rather than, say, a preoccupation with trying to save your soul, you may find that there *is* an atmosphere. I think I know what he was getting at, and I think it may have been truer then than it is now, for he had long conversations with natives who worked from small boats, and their preoccupations would have been very similar then to what he might have found on any small, inhabited, low-lying Hebridean island. Benbecula, say, or Gigha. Or – yes – Iona. But wandering round to the east coast of the island, stopping often and addictively to watch illimitable birds, from sparring pairs to great shoals of silvery-and-shadowy waders locked in a three-way dance with sunlight and sea, and some birds whose migrations have just ended and others that paused *en passant* on their way to Scandinavia or simply Scotland…on such a coastal journey I fancy I have contrived an atmosphere with a different accent. The feel does not strike me as that of Scotland's Hebridean coast so much as a singularly undiluted savour of its east coast. This is a fragment of East Lothian or North Fife or Angus, say, one that every few hours slips its moorings and goes for a paddle in the North Sea.

If you have the time to stay on Lindisfarne through a few ebbs and flows of the tide, you quickly fall under the seduction of the security of the raised drawbridge when the day's visiting flotsam has scurried landwards back across the causeway, and perhaps paused at the far end to watch it vanish before barely believing eyes. I do not often reach for Sir Walter Scott, but no 21st-century pilgrim or unambitious tourist to Lindisfarne can fail to be impressed by his unerring reading in *Marmion* of the particular spectacle they had paused to witness:

The tide did now its flood-mark gain.
And girdled in the saint's domain:
For, with the flow and ebb, its style
Varies from continent to isle.
Dry-shod, o'er sands, twice every day,
The pilgrims to the shrine find way;
Twice every day, the waves efface
Of staves and sandalled feet the trace.

The islanders, both human and otherwise, breathe easier once the waves efface the trace of staves and sandalled feet, not to mention tyre tracks, and the modest parcel of land that constitutes Lindisfarne is better able to sustain the natives and their guests who have come to stay for a while. I have in my back catalogue of visits an east-coast evening watching the moon heave hugely out of the sea and lay a second causeway of light across the waves even as they thundered ashore. It is then, when the horizon is a crumpled line of advancing waves, that an unshakable concept takes root in your mind and will never, ever let you forget it for as long as you continue to be an occasional visitor to the island. It is that you stand on a piece of land that is no metres at all above sea level. Given the concern that swirls around the more thoughtful science communities of planet Earth about rising sea levels and extreme weather events triggered by global warming, the consequences for Lindisfarne could be considerable. At the very least, it is surely a matter of time and nothing else at all before today's model of the causeway and the pedestrian option preferred by 1,500 years of pilgrimage are simply overtaken by events.

The islanders are reluctant to create a more permanent link to the rest of the world, and who can blame them? At its summer peak the tourist trade is utterly uncaring and somewhere beyond saturation point already. Perhaps the precious island atmosphere – and even those Hebridean traits Seton Gordon encountered but which have become more and more elusive – is ultimately retrievable not by a permanent link but with that most Hebridean of lifelines, a ferry, with controllable sailing hours.

A moonlit evening then, a spring tide of boisterous breakers, and the relative sanctuary offered by the uncrossable causeway, a shoreline more in thrall to the sea itself than most, is when you might expect nature to beguile you from the sea too. Instead, the natives emerged from the eerily moonlit and heaped-up landscape of the dunes. The shadows between the heaps coughed up perhaps the last thing you might expect to go hunting here – a family of foxes. With the low-tide flats out of reach for the moment, the beach was thick with birds, standing, sitting, dozing, waiting.

It makes no difference to the eyesight of a fox whether high tide is sunlit or moonlit or neither, but darkness tilts the odds inexorably towards predator and away from prey. It became clear almost at once there was a strategy, that it had to do with one adult fox making a conspicuous presence of itself while the other lay as flattened and stretched and still as a length of beached driftwood. Two leggy cubs sat at the base of dunes, engrossed, watching, learning. I stood in a shadow, engrossed, watching and learning myself.

The conspicuous fox trotted towards the waves as if it had not noticed the birds at all. Then it stopped, sat, looked around and did nothing whatsoever. A few minutes of this unnerved the birds. Finally, they snapped and panicked and flew, only to find as they took flight that the other fox was among them. I have seen this or something like it happen three times now, and on only one of the three was the tactic successful. The surprise is that the birds fall for it at all; you would think that evolution, race memory, whatever, would teach them that when the fox appears they vacate the premises. On the other hand, they work on the safety-in-numbers principle. There are hundreds in the flock, the worst that can happen is that two of their number succumb. So the odds of survival of any one individual bird are high.

Lindisfarne calls itself Holy Island, and has done for a long, long time. But really, it's no holier than any other. First and last, it is nature's island. The creatures of nature massively outnumber the human population, even when it is swollen by the unholy summer processions across the causeway and the holy ones across the sands. The oldest pilgrimages here are all nature's. We call it migration.

Chapter Eleven

May in June

THE COUNTY OF FIFE is no one's idea of wilderness. Much of it is quietly agricultural. There are small, Lowland hills and planted woods, a handful of bird-rich lochs enlivened by the occasional spellbinding intervention of a sea eagle. The south-west is still trying to recover from its long-dead coal-mining past. But its long coast, which binds the Firth of Tay in the north to the Firth of Forth in the south, is a simple, blunt-edged, wide-open collision between the rock of the land and the North Sea. There is no Hebridean atmosphere here. You look out on a sea without islands.

Except one.

Five miles out into the North Sea, off a bonnie thrust of land called the East Neuk, lies a scrap of singularly uncompromising terrain, the basalt love-child of the east coast sea haar and the east wind, just over a mile long and less than half as wide, and on an early June late morning it was drenched in sunlight, seabirds and the white blossom of sea campion as dense and deep as snowdrifts. The Isle of May in June is quite something.

From the mainland shore at Anstruther, where they have succeeded in elevating fish and chips to a high art and failed utterly to stop day trippers from tossing bits of fish and bits of chips to the gulls, the day was what you might call beautiful, the harbour tranquil, the sea as blue as alpine gentians on a postcard from the Alps. But when I called at the office of the *May Princess* to collect my ticket, the skipper's brow was furrowed. It was possible, he said, that we might not be able to land because of the swell out there. And because the entrance into the Isle of May's slipway is tight on a good day, there was a degree of concern. A phrase I learned on the Hebridean island of Mingulay came into my mind then – "not so much a slipway, more occasionally negotiable rocks". It hardly seemed appropriate to mention it at that moment. So passengers were being offered a refund, just in case they didn't fancy it. The skipper said he was sailing in expectation of landing, it was an improving picture, and it was simply a question of how much improvement and how quickly it improved. But this would be my fourth crossing to the May, the first three had been truly memorable, and the island-addicted acquire a feeling about skippers and crews, and these guys were good. I took the ticket.

The boat was full, about eighty passengers, and no one seemed to want a refund. Later, one of the crew explained that out there, "that bloody east wind" that brought us the Met Office's greatest hit of 2018, the "Beast from the East", had simply never stopped blowing more than two months later, and that explained the swell. So the skipper plotted a course that would enable us to turn straight into the swell, "which will be more comfortable for you". We all liked the sound of that word, "comfortable". We were not all sure we believed it was appropriate, but when it comes to small ferries in the North Sea, "comfort" is a relative commodity.

Puffins in ones and twos and fifties and sixties, and gannets in wavetop fives and tens, streamed across our bows almost as soon as we were clear of Anstruther harbour. Cormorants and shags, guillemots and razorbills, sat on the swell (looking smug, mocking our caution), and grey seals eyed us uncertainly and stroked their whiskers. On board, we amounted to a motley ballast. This was the place and this was the season for an impressive array of "mine's bigger than yours" telephoto lenses, the biggest of them like steroid-enhanced baseball bats. Like the binocular-flaunting scruff on the Colonsay ferry, there was a certain amount of ostentation going on, a surprising number of passengers were wearing a surprising number of thousands of pounds worth of gear around their necks. And a certain amount of good-natured geek-photography conversation drifted across the deck. Others pointed their phones. The accents were multi-national, multi-lingual. But "puffin" is easy in any language. And everyone laughs and smiles at puffins. And the sun bore down on us and all was well in our world, and the island advanced down the sea towards us and its cliffs grew tall and acquired detail and shouted "kittiwake, kittiwake" at us. The *May Princess* danced to the sea's tune and our spirits were high, until it came to…

The tricky part.

The skipper homed us in on the entrance between occasionally negotiable rocks. That entrance seethed with white water. The undulations of that troublesome swell were, admittedly, well-spaced, which also meant we had time to see them coming. They looked muscle-bound and, well…big. In fact, BIG. One by one they toyed with the *May Princess*, had their fun, and moved on, then they thudded into the island with a percussive roar, and the waters of the entrance became temporarily not-negotiable again. And again, and again, and again. We bounced, slewed, waited, held on to something, waited. The skipper in the wheelhouse watched the entrance, turned to look over his shoulder, while one of the crew appraised each wave from outside the wheelhouse, called out the possibilities or lack of them. Finally, there was a moment, collectively and individually we girded our loins, and we went in. Suddenly the water was calm, glassy, the waves elsewhere. We, the passengers, burst into spontaneous applause that was part admiration, part gratitude.

At once, a hundred Arctic terns swarmed into the air above the inlet, and the rocks amplified and threw their voices at us. It was never the most musical of bird voices, but in the same way as unexceptional human voices can sound sensational in a choir, the effect was of both a glorious anthem of nature and a spectacle of extraordinary beauty.

And this was why we came. This was what we were here for. We stepped ashore to make our own accommodation with the Isle of May. We would worry about getting out again when the time came. From first moment to last, and throughout the island, this glorious ritual of the terns and its soundtrack rose spontaneously, as if it were the voice of the island itself and that it might have some message to communicate to us. Every time it happened, it stopped me in my tracks. I listened hard. I stared. I was, frankly, thrilled by it. It struck tremors of who-knows-what between my shoulder blades. I felt grateful, almost worshipful.

The backward spring had held the terns back. They were late in arriving, like so much else, and their nesting overtures had just begun. Once the birds are sitting tight, you don't get these extraordinary flypasts. Walking on the path through the nesting grounds is not like this. Instead, they pick you off one at a time and fly at your head. If you are lucky, they brush your hair with a wingtip; unlucky, and they draw blood. If you have been here before and know the ropes, you raise a hand high above your head with a camera in it and you press the shutter button and keep pressing as the terns take aim. There is always the chance of a good photograph (in my case, one in about 200 attempts among the tern colonies of my life), and they won't draw blood because they aim for the highest point of any one human interloper. And if you are on the Isle of May at nesting time and you enjoy free entertainment, it's worth getting back to the boat twenty minutes before sailing time, finding a good spot on the upper deck, and watching your fellow passengers take evasive action as they run the gauntlet. You will laugh yourself sore.

My particular comeuppance on that score was two years ago when, being well schooled in the ways of Arctic terns, I deployed my usual high-camera-hand technique and escaped unscathed, only to be ambushed moments later by a greater black-backed gull, which is about the size of a Lancaster bomber to the tern's Spitfire. I still bear the scar. There is a moral in there somewhere, and I am still trying to work out what it is.

Ah, but then there are the eider ducks.

By comparison with the terns, there is no more stoical, immovable, silent nester than the eider duck. She looks you in the eye and just sits there on her sumptuous couch plucked from the down of her own breast, the living, breathing definition of "dead still". She is a yard from the path. She has felt the earth move beneath her as she has watched the footwear of a boatful of us pass by, some stopping right beside her to point cameras and phones, others inexplicably unaware that she is there, just a yard from their footfalls, fooled by stillness and the best natural camouflage on the face of the island.

A curious truth falls into place here. It is that although these must be among the most visited – and therefore the most disturbed – seabird acres anywhere in Scotland, they are also among the most successful. The island is a National Nature Reserve, owned and managed by Scottish Natural Heritage, access is by ferry only, and from Easter until the first of October. And because visitors know this, and because they come to see the birds, they also come in an attitude of respect, and their behaviour is almost universally predictable. Despite their blatant scrutiny and the chatter and general craziness that emanates from around eighty widely scattered humans constantly on the move, the birds thrive. Because of the peculiar circumstances of the Isle of May, and a regime that insists that everything is secondary to the well-being of the birds, this benevolent ritual of people walking only the network of paths actually appears to function like an extra protective membrane. The birds are safe from everything but each other. (A black-back swallowing an eider chick whole is as commonplace and as distressing for a human observer on the Isle of May as it is anywhere else.)

The eider colonies are a case in point. All around the country their numbers are in free-fall, but here they nest and they hatch and they fledge in record numbers. A survey in May 2018 revealed 1,183 nests, an increase of four per cent since the previous survey in 2016. Consider the nature of spring 2018, and be very, very impressed.

Up in the centre of the island there was a pool with a ditch. Just where the ditch enters the pool, two eider ducks were escorting six newly hatched ducklings on what is the defining rite of passage of every eider ever born: wherever the nest, they have to get to the sea, for there, and only there, lies relative safety. That means running the gauntlet of predators, and on the Isle of May, that usually means the big gulls. The sea means the safety of crèches, amalgamations of broods well marshalled by several females. I watched the six blackish-brown chicks plod up a short patch of mud and disappear into the ditch with one duck in front and the other behind, riding shotgun. She kept looking back, but the direction of travel was resolute. I wished them safe passage. It's a tricky and imperfect art, raising a brood of eiders once the nest is abandoned, but they work on the basis of safety in numbers, and on the Isle of May in particular, they get the job done.

Trickier still is managing landscape for nature, and as a species, we are not very good at it. Many times in various publications I have quoted New Hampshire wildlife writer and painter of distinction, David M. Carroll: "Let wildlife manage wildlife." It is the one solution that never seems to occur to land managers in Scotland, even in the more adventurous conservation organisations. And while Scottish

Natural Heritage could never be called adventurous, it seems to me that something like that is happening on the Isle of May, and the results are impressive.

◉ ◉ ◉

That island day of birds in never-ending raucous splurges set against a huge, blue east-coast sky, or against breeze-bowed, meadow-like drifts of sea campion, finally distilled down to two individuals, one eider duck and one puffin. These caught my admiring glance and held it, and kept returning it again and again, so that the memory of them has become indelible. I have been studying two photographs, remembering, reconstructing their extraordinary moments.

The eider duck was sitting on a nest, I think. The doubt stems from the lie of the land, for it masked the entire bird apart from the head, which appeared to me as a left-facing profile. Sunlight lit the top of the head, the forehead, and the upper mandible of the bill, the top of one eyelid, the top of one cheek. The back of the head, the whole of its neck and the lower mandible were in deep shadow and looked black. Every square centimetre of the picture was crammed with swarming sea campion. The head of the bird was perfectly positioned at the base of a shallow vee in the land that extended from one side of the picture to the other. The edge of the vee was also where the sea campion flowers were in sharp focus. A tiny cluster of seven flower heads was particularly striking because its immediate background was the black-looking mass of the eider's neck. The bird must have been sitting on a ledge, just beyond the vee and hidden by it, which was why only the head was visible. The top half of the picture was the high bank that rose some distance behind the bird and beyond the vee, and it was an abstracted, out-of-focus mass of green and white. From top to bottom and edge to edge of the picture, the only thing not sea campion in various states of sharpness was the head of the eider duck, and even that had lost the tip of its bill because it was masked by flowers. The stillness was somehow unnerving. So was the unshakable confidence of the bird in the sureness of its place on the map, its place in the island's scheme of things. It impressed me. Not least because such a sensibility is quite absent from my own life. My long-standing admiration for eider ducks just deepened perceptibly.

The puffin was also on an edge. But unlike the eider's fragment of the island, it was a cliff edge, and I could see the whole bird. It stood, presenting a right-facing profile, and it looked at me looking at it, and we were about thirty yards apart. After a couple of minutes of this, it flew, and I assumed my audience was over. But it was back a minute later and came into land with orange webs thrust forward and whirring wings held high, to land in precisely the same spot, to assume precisely the same attitude as before. This happened five or six times over quarter of an hour. Each time it flew, it couldn't have travelled any distance worth speaking of, did not land on the sea or dive beneath for food. I wondered if it was a game, or if it was unpaired and looking for a mate and that my presence and stillness was completely incidental to the ritual.

But on the sixth or seventh occasion I noticed something I had never seen in puffins before – and from Shetland to St Kilda, the Treshnish Islands to the Farne Islands, and

from Auchmithie on the Angus coast to St Abbs on the
Berwickshire coast and the Farnes off the Islandshire coast,
I have seen a *lot* of puffins. So when this Isle of May puffin
returned yet again just after I saw what I had never seen
before, then stood facing right and hung around for a while,
I was ready for when it took off again, to see if it repeated
the manoeuvre. As soon as it turned its back to face the
cliff edge, I focussed my camera lens and pressed the shutter
button as the bird prepared to launch into space and kept
pressing. I took one useful photograph. It showed the bird
the instant before it left the cliff edge. The head was forward, the back was almost horizontal, almost parallel with
the ground. The wings extended horizontally from the
"shoulders", but only as far as the outer end of the "fore-
arms", where both wings were angled down well past ninety
degrees so that the wingtips pointed back *inwards* towards
the bird's feet. It was in that attitude that it became airborne.

Then the bird was gone, the cliff edge was empty apart
from a frieze of sea campion, and I was scratching my
head. What was that manoeuvre designed to achieve? Air
brakes? A tight bit of precise steering round the back of an
inland-facing cliff and through a gap beyond before it could
reach the open sea? Or extra speed? Eagles accelerate by
half-folding their wings, but not like that. I consulted the
one book I own that analyses bird flight and explains why
certain birds deploy certain techniques to meet different
circumstances. None of them approximated to what I had
just seen. My dilemma was compounded by the fact that the
puffin had vanished over the edge the instant it took flight,
so I had no idea what followed. Perhaps I could pin down

an expert in the sophistication of bird flight, and all would be prosaically explained. In the meantime, the extraordinary stillness of the eider, the restlessness of the puffin, and my constant interaction with the raw stuff of the island and its sea (and the mainland shore that set both off to such good advantage), have left me with a very persuasive sense of wilderness ensnared in a scrap of basalt anchored five miles off the coast of Lowland Fife.

Part Three

Highland Spring

Chapter Twelve

Glen Clova and
the Definite Article

GLEN CLOVA IN THE county of Angus is the native heath of
my mountain heart. I have made the flimsiest of scratches
on the surface of other mountain lands – Alaska, Iceland,
Norway, Switzerland – and I have a store of memorable
away days to England's Lake District, but Highland Scotland
has always been mountain enough for me, all the mountain I would ever really need. I have lived most of my life
in sight of its southmost mountains, the exception being a
spell of fifteen years when I lived *among* those southmost
mountains, in Balquhidder and Glen Dochart. One way or
another, crossing the threshold that is the Highland Edge,
that frontier between Highland and Lowland, is second
nature to me and sometimes it can feel like first nature.

I was born in Dundee on the north shore of the Tay
estuary, and, historically at least, it too was in the county
of Angus. Dundee is built on two hills, the Law and the
Balgay. I lived the first twenty or so years of my life on the
slopes of the Balgay, and with a view that included the Law.
The Law is the summit of the city, the centrepiece. From
up there and on the right kind of day, you can see the bulky
mountain upthrust of The Dreish in the Angus Glens, and

all the way from there to the elegance of Schiehallion in Perthshire, and these were my childhood symbols of the promised land. Between the city and the Angus Glens lie the Sidlaws, the lowly and lovely bulwark where I served my apprenticeship and learned to fend for myself in the company of hills; to the north of the Sidlaws lie the Angus Glens, and cradled in the midst of the Angus Glens lies Glen Clova, and Glen Clova is the home of The Dreish. So Glen Clova is the native heath of my mountain heart.

When the Sidlaws were still the northern frontier of solitary teenage explorations, I would eavesdrop on the conversations of old boys who frequented the glens, relished and envied in equal measure the way the mountain names tripped off their tongues with casual familiarity, as if they were friends – Mayar, Craig Mellon, Broad Cairn and (improbably exotic and umbilically linked to the Angus Glens) Lochnagar. But the one that intrigued me more than all the others was the one that had not just a name but a definite article. Maps may disagree, but to all of us who grew up into that mercifully benevolent Angus hill-going tradition, it will always be *The* Dreish. And it was there, five years after he had died in 1973, that I heard my father's voice utter the four words he reserved for any hill view he ever admired anywhere in the land:

"Ah, the bonnie hills."

And forty years after that, in the backward spring of 2018, driving to Glen Doll through Glen Clova for a walk organised by *The Scots Magazine*, to which I contribute a monthly column of nature writing, I suddenly remembered the moment again. The trouble with travelling to Glen Doll is having to drive through Glen Clova without stopping. It's a bit like having to drive through Skye simply to catch the ferry to Harris, except that in Clova, the roots go deeper, and not stopping is more of an issue than you might think. Clova brings out the contemplative in me, and I have come to think of it as the place I go to after long intervals to reacquaint myself with my sense of who I am and what I stand for. So as I drove through Clova that winter-into-spring day, I resolved that as soon as I had time I would make a space in my life and re-establish a clear perspective between myself and these mountains.

These were where I took my first mountain footsteps. Whenever I haunt them again, I have to acknowledge that between Glen Clova and Glen Doll and that fork of peaks from The Dreish to Broad Cairn and Craig Mellon, all their corries and lochs and passes, and that miles-wide offshoot that trundles under the sky and over the Capel Mounth… all that country pushes me deeper into myself than elsewhere. And The Dreish itself was not just the first mountain I ever climbed, it was the first I ever saw, the first that ever frightened me, the first I ever slept out on. If I were ever to hear my father's voice apparently addressing me directly years after his death, it would have had to be here. And from time to time, it just feels good to go back.

So I took the Capel track on a mid-March morning with the mountains braided with old snow; the Capel track because of the way The Dreish rises and expands across the glen as you climb and turn to look over your shoulder. A squall thrashed down the glen out of the north beneath a grey-black cloud that was nevertheless sunlit along it its lower edge. Curtains of rain slanted southwards as the squall travelled, but at the same time the cloud began to burst open and beams of sunlight slanted northwards, so that the whole glen, or as much of it as one pair of eyes could take in at any one time, was strewn with diagonals. For a moment there was a girder-like image in the midst of it all that might have been a cross section of the Forth Bridge. The immediate foreground of the Capel track's hillside was dark, the floor of the glen was lit bright green, an uncanny shade, the long ridge of The Dreish immediately above the glen was studded with those snow patches, and again I thought of Cézanne and those flecks of unpainted canvas on his mountain paintings. The effect of the squall and the contrary sunlight was to soften and inflate the mountain so that it looked immense.

The light in the east of Scotland is a phenomenon unknown in the west; it washes ashore from the North Sea and pounds inland over fields and foothills and reaches the mountains around Clova in twenty unobstructed miles. When Joseph McKenzie established the photography department at Dundee's Duncan of Jordanstone College of Art in 1964, he called that light "pole-axing". The mountains of the west crowd round the coast and wade towards it, and are saturated by sunset light. But here, the sun rises out of the sea and fires up furnaces of light there that flare across the sea and barge inshore. My most enduring single image of my young self growing up in Dundee is of paper-round early mornings on my bike in spring and summer, stopping to watch the sun rise out of the sea. It just takes a moment like that one in Glen Clova looking across at The Dreish and the phenomenon of the light show in the intervening space to crystallise and compress my own life in my own landscape, and to reinforce in me the great good fortune that befell me growing up where and when I did.

The air is part of the mountain, which does not come to an end with its rock and its soil. It has its own air; and it is to the quality of its air that is due the endless diversity of its colourings.

Nan Shepherd wrote that in her 1977 lyrical hymn to the Cairngorms, *The Living Mountain.* I can imagine the ghost of Paul Cézanne reading that, grunting his approval, and muttering:

"There, see? I told you so. She knows."

Then, as I climbed on, Craig Mellon locked itself into its place in the landscape, and I turned again at the very moment when a slim shaft of sunlight danced along the ridge from The Dreish and arrived at the point where Craig Mellon wore a perfect down-curving rim of snow, as slender and tapering as a curlew bill, but about a hundred yards long.

This is my way of climbing mountains when I'm alone; this is why I am not good with big groups, because it feels as if I have to submerge too much of myself in deference to tribal loyalties, to conversation, to the even progress of the group. My instinct, when it is unfettered, is to stand and stare, to question, to write down, to draw, occasionally to photograph (though increasingly I find the process intrusive), to take the time to marvel, to relish, to be a fragment of landscape, to travel like one of Louis MacNeice's "Great black birds that fly alone / Slowly through a land of stone".

As the track climbed and the mountain landscape around me slipped into its old familiar pattern, like a gathering of old friends, I stopped looking over my shoulder at last and began to look forward, because just up ahead was that moment when a high moorland crest falls away and Lochnagar stands there like a mountain in a dream. Every time, it astounds me. Every time I know it is about to happen, and every time I am quite unprepared for it. But the first time, alone and unwarned, the mountain was embattled in its own rage of storms. Only Meikle Pap stood clear of the storm like a dark lighthouse warning me not to approach too close. I stuck to the track that day and reached Loch Muick in a kind of turbulent daydream.

There is something to be said for revisiting landscapes that have featured as defining moments in a lifetime, something of value in savouring the moment when they first exploded before your eyes, when they blew apart your preconceptions, long before familiarity would change the nature of the relationship, long before you have learned to love a particular landscape. That love is born of many visits, but there is only ever one first time, beyond which you are never the same, because your native landscape has taken you over a threshold beyond which lies a new land of possibilities. A few years later, climbing up on to the Moine Mhor from Glenfeshie in the Cairngorms, I reached that point, that nowhere-at-all on the map where you take one more step across the plateau and suddenly find Cairn Toul sprawled across your horizon. Two things happened in the same moment: I gasped out loud, and my mind lurched back to Lochnagar and the Capel Mounth. Only the scale was different. The sense of revelation was the same.

But now I was back on the Capel track, and I was ready for the revelation of Lochnagar, and my head was full of all the once-upon-a-times that had unfolded here. The name that falls into place hereabouts is that of Syd Scroggie, Dundee's very own mountaineering legend. Syd was a friend of my father and later in his life he was mine too. He was legendary because, during the Italian Campaign in the last few months of World War II while serving in the ski-mountaineering regiment of the Lovat Scouts, he encountered an anti-personnel mine that left him minus a leg and blind. But Syd was made of stronger stuff than the landmine because with an artificial leg and a helping hand, he returned to the mountains and climbed them again, and when a memorial cairn was unveiled on Baludderon Hill in the Sidlaws in 2000, he was there, aged eighty. He finally died six years later. I have a handful of his meticulously typed letters from between 1987 and 1999.

"Clova is really my spiritual home," he wrote in one, and in another:

...that glen, which quite as much as it is yours, early took possession of my soul as satisfying its deepest requirements. In my case, it's something to do with the smell of woodsmoke, paraffin lamps, Orion as seen on a frosty night between Craig Mellon and Craig Rennet, mist-shrouded peat hags on Jock's Road, fag-smoke in Sandy Hillock's hut, dawn on the top of Broadcairn, and on a dreich glen day Ma Harper saying at the door of Newbigging, "I aye ken it's goin' tae rain when the whaup cries yon mournfu' wey."

I had obviously told Syd about hearing my father's voice on The Dreish, for in another letter, there's this:

I can now formulate a theory to account for a grafitto which has recently made its appearance on the Dreish's trigonometrical pillar. "Hello," this simply states, and perhaps it wis pit there by auld Jimmy for the benefit of his blinn'd freend. We did the Dreish a couple of times last year, Margaret and I, and it's an eerie thought, if think about it I must, that it's 50 years now since I got up this hill first.

Tempus will *fugit*, Syd, and it can't be far short of fifty years now since I got up this hill first myself. And here I was on the Capel track with only the sound of my own feet for company, forby my thoughts of Dad and Syd, and a rhythmic snatch of one of Syd's poems got stuck in my head:

I will attempt the Capel track
Old, stiff and retrograde

The lines had fastened to each other at both ends so that they came round again and again like a conveyor belt, and although it's a fine opening to a fine poem, there is a limit to how long you can admire two lines of anything without craving the arrival of the third line. The last crest grew close, so I hurried forward to break my stride pattern and let the poem defer to the moment of intense anticipation when Lochnagar would hove into view. Or not.

I could see where it should be, but it was an amorphous mass of black, a shroud, unrecognisable, stormbound, lost.

Yet no one ever crossed that moorland crest empty-handed, and instead of Lochnagar, Connachcraig stood clear of that storm; the moor itself was pale gold, and stitching the moor to the hill in a low-slung curve was a rainbow. Nature was playing scene-shifter and had replaced the mountain with the rainbow. And suddenly, as I stood stupefied at the wonder of it, Syd's verse fell into place, and I spoke it aloud to nobody at all:

I will attempt the Capel track
Old, stiff and retrograde
And get some pal to shove me on
Should resolution fade,
For I must see black Meikle Pap
Against a starry sky
And watch the dawn from Lochnagar
Once more before I die.

All the way to Loch Muick the rainbow stayed before me, while the base of Lochnagar glowered bleakly at the moor. I went breezily north, pleased I was having the high moor of the day rather than the high mountain of it. Wind and rain stalked the track with me, but it was light, airy stuff and it was invigorated with sunlight. My season was spring, the mountain's still direst winter. I headed for the Black Hill, the perfect mountain-watcher's stance from which to admire the spectacle of Lochnagar on days when there is any spectacle to be admired. I collect such places, for the one place from which you cannot admire the shape and the stance of a great mountain is on the mountain itself. So I number among my beaten paths such unsung places as the Black Hill; Beinn a'Chrulaiste, which contemplates Buachaille Etive Mor; Carn Elrig, for its side-long glance into the Lairig Ghru and its head-on, two-eyed view of Braigh Riabhach; the tilted wedding cake of Dun Mor, for the way it stares into the Prison of Skye's Quiraing; and Beinn Bhreac on Soay, the ultimate Cuillin-watcher's pedestal.

The Black Hill rises no more than 500 feet above the moor, but that is more than enough, and perfect for watching Lochnagar, Broad Cairn and that long south-making horseshoe of mountains as far as Mayar. Stay out all night in late spring or earliest summer and win the rare privilege of seeing them lit by that sea-born sunrise in the north-east. Curlew and golden plover might serenade you through a darkless night if you are as lucky as I was, how long ago now…thirty-five, forty years? Lochnagar burned while Mount Keen smouldered and the Black Hill itself glowed like hot coals. A perfect collar of cloud hung around the neck of Meikle Pap, smoke-grey until the sun touched it, and over five minutes the thing dematerialised in fragments of fire. The rock stood unadorned and deep dark red, then sunlight flooded the whole mountain mass, and I remember nothing at all of the way down and the drive home.

Syd's poem, which is called "Ante Mortem", has a second verse, offered here in heartfelt tribute to the wholly admirable spirit of a poet-mountaineer who is himself a part of the fortunate inheritance Dundee folk call "goin' up the Clovy".

And if I do not make the top
Then sit me on a stone,

Some lichen'd rock amongst the screes
And leave me there alone,
Yes leave me there alone to hear
Where spout and buttress are
The breeze that stirs the little loch
On silent Lochnagar.

Chapter Thirteen

The Poetry of
Mountain Flowers

EVERY APRIL, I MAKE the same promise to myself: I must become more fluent in the poetry of mountain flowers. I don't mean flower poetry ("Wee, modest, crimson-tipped flower"…"I wandered lonely as a cloud"…"Cwa' een like milk-wort and bog-cotton hair"…that kind of thing), but rather the almost invariably accidental poetry that characterises the language of the names of mountain flowers, whether composed by scientists or by real people.

This annual rekindling of good intentions is mostly occasioned by my first encounter with the mountains' idea of spring, with purple saxifrage. "*Saxifraga oppositifolia,*" purrs the botanist, mysteriously scorning the bunched fists of startling purple flowers in favour of the plant's prosaic (some would say boring) character trait of leaves in opposite pairs, hence *oppositifolia*. Such dimwit tendencies make me wonder if science sometimes mislays its brains, yet no one with an ear for rhythm and cadence can deny the buoyant, flowing surge of *Saxifraga oppositifolia* when you roll it around your tongue.

The Gaelic, however, is something else: *clach-bhriseach purpaidh* – "purple stone-breaker". What genius came up with that one? What inspired mountain wanderer first examined an eye-level cluster of purple saxifrage under a dripping-wet rock overhang that far-off spring day, noticed that it was rooted in the very rock itself, and reached for the image of a stone-breaker, as if the plant had prised the rock apart by virtue of its own muscle-power, rather than parasitically exploiting an existing fissure? Whoever it was, a poet's sensibilities were at work. Anyone who believes that you cannot get blood out of a stone has never seen the *clach-bhriseach purpaidh* flex its tiny muscles on a wall of old granite. Likewise the sceptic who thinks faith cannot move mountains.

The poetry of saxifrages is even better served in the case of the starry saxifrage. In truth, it is no starrier than any of our other native saxifrages, but the sound of its Latin classification, *Saxifraga stellaris*, glitters like the nebula of Orion. And the Gael's *clach-bhriseach reultach*, the starry stone-breaker, is surely worthy of a constellation all of its own. The starry saxifrage and I go back a long way. I first met one face-to-face high on Beinn a'Bheithir above Ballachulish, and so long ago that it cost me a shilling (I was *very* young at the time). I had pulled a hanky (another artefact from a bygone age) from my pocket and the coin tumbled out with it, bounced once and disappeared into a crack a few inches wide. I fancied I could still see it glinting palely there, and I crouched down and peered into the gap. But all was black beneath except for the upturned star of a single white flower. It was starlight I had seen, not a shilling. Not long before this encounter, I had acquired the invaluable knowledge of inverting my binoculars to turn them into a microscope. This was its first practical application, and I discovered that the starry stone-breaker is much more than a five-pointed white star. Firstly, there are twice as many tiny red stamens as there are petals, and secondly, there are two vivid yellow spots on every petal. I know now that this discovery has the kind of impact on the botanic community that pointing out the red breast of robins has on ornithology, but then and there I was on my knees in wonderland.

And what really moved me was the tiny daring of the thing. The crack opened into a black underworld two or three feet deep and about as wide. There was space there for a whole Milky Way of starry saxifrages, but apparently sustenance enough for only one. That single stem with its single flower head aglow in that chamber of gloom is surely the most primitive – and the most poetic – thing I have ever seen. I have tried several times over the years to make a poem of it, and I will probably try again, but so far the thing has proved too elusive for me, as elusive as a shilling lost in a hole in the ground.

If none of that impresses you, perhaps I can interest you in the case of the mountain avens. The late glacial period of our last ice age, about 10,000–15,000 years ago, is known among people who worry about such things as the Dryas period. "Dryas" may not be the most lyrical-sounding of names, and *Dryas octopetala* is more percussive than lyrical, but there is a kind of poetic rightness about its unique significance. It is better known as the alpine flower, mountain avens. *Octopetala* is fairly straightforward – eight petals – but why *Dryas*? I'm glad you asked. Such was the profusion of mountain avens fossils datable from the late glacial period that the people who make decisions about how to christen swathes of time lasting 5,000 years drew their inspiration from arguably the loveliest of alpine flowers. No flower was ever better designed to make its mark on an ice age, or, for that matter, to thrive in 21st-century Scotland. From the north coast to Kintyre, and from Angus to the Hebrides, it turns the white star of its fair face to follow the sun through the day, likes cool and humid exposed places and has a contemptuous attitude to frost. So it is right at home on the Atlantic coast at sea level and on the flanks and broad ridges of our Highland and Border hills.

Alas for the mountain avens, poetry has rather let it down, which is strange and a touch unjust considering its beauty. Even the choice of *octopetala* is as suspect as it is prosaic, because out there in the wild world it can have anything between seven and ten petals. Perhaps science took its cue from the Gaels, whose *machall monaidh* means something like "large-flowered mountain plant", whereas any self-respecting poet would have mustered something like "yellow-hearted star of the mountain snows".

The older books of Seton Gordon offer intriguing glimpses into that inches-high world of alpines. Here he is on the subject of dwarf cornel, also known as *Cornus suecica*, a favourite of many a Cairngorms walker:

> ...At first glance, four white or pale yellow petals tinged with purple appear to grow on each flower stalk, but a closer inspection reveals that these "petals" are in reality "bracts" and their function is to attract by their conspicuous appearance the attention of the insects that fertilize the plant. The bracts surround a large number of minute purple flowers with yellow stamens.

The fruit is a red berry. It was said by the Highlanders to create appetite, and the plant was called by them lus a' chraois *– the herb of gluttony…*

Poetic or what?

It is strange, now that I think about it, that I haven't tried to write more flower poems (two, actually), especially given the amount of time I spend on shores and mountains in thrall to those tiny species that seem to be able to move effortlessly between such extremes of habitat, but shun the in-between zones; something like sea pinks, for example, also known as thrift, or even sea thrift. They dance on the edge of the high tide, blasted by ocean winds and half drowned by salt spray, yet they also have the extraordinary capacity to smother the highest plateau thrusts of the Cairngorms at almost 4,000 feet in fleeting magic carpets of pink. Sometimes in its mountain guise it flowers for no more than a midsummer week or two before it is plagued by the Cairngorms' appetite for midsummer frosts or blizzards, or both, and shrivels to blackened husks.

Seton Gordon also wrote this little aside about it:

*…Most botanists believe the thrift of the Cairngorms to be the sea thrift (*Armeria maritima*). An article by Mr Harold Stuart Thompson, F.L.S., on the thrift of the Scottish hills appeared in* The Journal of Botany *in 1910. This writer thought that the thrift of the hills was a form of* Armeria maritima, *but Mr F.N. Williams has suggested that it is really* Armeria alpina, *the thrift which grows on the western Alps from 7,000 to 9,000 feet above the sea.*

Posterity seems to agree with Mr Thompson, and so do the Gaels, who call it *neomean cladaich* – "daisy of the shore" – wherever it happens to grow. Anyway, in the spirit of the poetry of mountain flowers, whether rooted on the Cairngorms plateau or on an island shore 4,000 feet lower, here is one of my two flower poems because it happens to be about thrift and because it's April again and I have just renewed my promise to myself.

Pink
Pink, you might think,
is a shade that ill-becomes
a Presbyterian sense of place
until a self-sown sea of thrift,
a drift of glowing snow,
invades the tidal space.

One Sunday-savvy otter,
respectful in Sabbatarian gray,
contemplative as a monk at Lent,
forgot herself that day,
and half-drunk on the scent,
squirmed wantonly, singing,
then trotted guiltlessly home
chasing butterflies and bees,
the tell-tale petals still clinging
to the fur above her knees.

I mentioned above that I have only written two flower poems. This is the second one (strange that they should both feature monks, but in case you are not well versed in jazz, the Monk referred to in this poem is Thelonious of that ilk, jazz pianist of distinction):

The Softly Blues
The softly blues of speedwell
by a shadow-and-sunlight mountain tree
are pianissimo, man,
Monk in a minor key.

Easy now
with only height to lose,
I got those low-on-the-mountain
softly speedwell blues

Chapter Fourteen

The Sanctuary (1) –
A Second Spring for the Wolf

THE TREES I HOLD dearest are those that remember wolves. You will find them, for example, among the Scots pines of the Black Wood of Rannoch. Some have great girth and vast reach, having demanded their own space some-time in the first fifty years of their youth, and more than 200 years ago. Some are older, but dead. After 300 years of life, one of these has already stood for a hundred years of death – leafless, barkless, broken, bare, a tree of bones. But death for such trees does not impair their remembrance of wolves, any more than it impairs their capacity to sustain uncountable generations of uncountable numbers of lesser lives, lesser than Scots pines, lesser than wolves.

Wolves are not so long gone as you might think. Sometimes, here and there amid heartland tracts of Highland Scotland, like the Black Wood or the Black Mount hills and woods around Loch Tulla, I have become aware of an absence, and I put it down to wolves. That sounds finicky, I know, but such landscapes have a way of dislocating time. There is a harder-edged wildness at work there, and the sense of it reaches you almost like a scent, or the sound of it is in your ears like a threnody, for it laments its own incompleteness. For thousands of years, this was a land shaped almost as much by wolves as by ice, and the wolves were shaped by the forest. And now, much of that forest is an absence too.

These are quiet places where a solitary traveller can still feel close to the land itself. In the far north of the world, to this day there is an awareness among circumpolar tribes of the need to *listen* to the land, a phenomenon I have raised in this book's two predecessors, in the context of the crisis of climate. But if we were to relearn that skill, to reach out to those who still teach it (and who learned it from wolves), and if we took that skill out into old wolf heartlands like these, we too would sense the restlessness in the land and become articulate in its ancient speech; we would learn that the land is ill-at-ease because what we like to call our "stewardship" has deprived nature of the essential tools it needs to perform at its utmost. I believe that the absence of wolves in a northern-hemisphere country like Scotland is the most glaring manifestation of that ill-at-ease condition. If you listen to the land today, if you sit still and quiet and alone in its midst, looking around you and cross-examining what you see and what your hear and what you feel, the land will tell you that it still feels the absence of wolves, that it still mourns wolves, that the memory of them is immortal.

When Scots use the words "immortal" and "memory" in harness, we are invariably referring to Robert Burns, that cornerstone of the Burns supper address known as the "Immortal Memory". He is as alive in our midst as he ever was. And it is quite possible that Robert Burns knew wolves. He would certainly have known wolf stories. His parents were born into a Scotland already overflowing with wolf legends and wolf lies and these included an obsession with "last wolf" stories, even as the dwindling remnants of Scotland's wolf population scattered to the unpeopled corners of the land, there to make one last stand for nature. By then, the land was already restless.

Wolves have been gone from Rannoch for more than 200 years, but perhaps not much more. It is a perfectly plausible proposition that they held out in Rannoch's twin fastnesses, the Moor and the Black Wood, for decades after they had been rendered extinct in much of the rest of the land by the people because…because they were wolves. Tomorrow, some sweet tomorrow of our own choosing (and the decision is ours to make and the process is simple and the difficulties more or less non-existent and ecological benefits immense and permanent), we will declare the place a Sanctuary, and they will return to Rannoch, and the remembrance of both tribes, wolf and Scots pine, will renew and refresh.

Why Rannoch? Practicality and symbolism; the required beauty, the required grandeur, the required scale; and yes, if I am honest, for a stage that befits the dreaming and daring of such an enterprise of wild theatre. But mostly, it's the natural choice for the practicality and the symbolism, for Rannoch is at the centre of the land, and (as we have seen from the Lismore chapter) the forge that fashioned the shape and the nature of our post ice-age land, radiating glaciers from its towering ice cap in every direction, furrowing the land with glens and flooding them with lochs and sea lochs and firths. Besides, travel is in the nature of wolves, and over subsequent decades, new generations can radiate outwards from the centre in whatever directions they choose, following the natural spoor laid down by the Rannoch ice cap and its glaciers. It is the most organic process imaginable. And as they radiate, they will spread the lifeblood of ecological renewal, for theirs is essentially a benevolent, reinvigorating regime, and a healing regime for nature.

And if there is to be a future for wildness in Scotland on a scale appropriate to the nature of the land (and surely there must be, for the alternative is a long and inexorable descent into that dark place that accommodates only the ugly places of the Earth), then there should be a showpiece at the heart of things, one within reach of all of us, so that we – *all* the people of Scotland – will be as neighbours in an enterprise where nature makes all the rules and all the decisions, establishes all the priorities. And this showpiece will belong only to nature itself, and to the nation of Scotland. This is what we have never dared before: to take back a portion of land from "private" landowners who put their own interests before nature's and indulge themselves – and only themselves – by slavish adherence to a code established by the Victorians, and for whom the reintroduction of wolves is akin to a pact with the Devil. That land will be set aside, we will restore and expand its native habitats as far as we are able, remove the alien traits of the legacy of the "sporting estate", then stand back, make a respectful space, let wildlife manage wildlife, and watch and learn.

Conservation has a 21st-century buzzword: "rewilding". We should be clear about one thing – despite the rhetoric of celebrity talking heads, *we* cannot re-wild anything.

Wildness is in nature's gift, not ours. All we can do is remove from nature's path the obstacles our species has put there, obstacles that diminish or deny wildness, obstacles that inhibit nature. Having removed them, what we can also do is replant and reintroduce so that nature has something to work with, and once we have met those of its essential requirements we are capable of providing, our job is to back off. Then, and only then, is when rewilding begins, once nature has that something to work with, and that something is a full set of tools. And Aldo Leopold wrote in *A Sand County Almanac* (Oxford University Press, 1949):

> *The practices we now call conservation are, to a large extent, local alleviations of biotic pain. They are necessary, but they must not be confused with cures.*

Scotland's landscape requires cures, nature requires us to allow it to flourish unimpeded and on a scale it has not enjoyed in a thousand years. It sounds complicated and daunting. It is neither. One ferociously simple idea can go a long way towards achieving it.

Restore the treeline.

The treeline is the upper limit of altitude at which native trees grow naturally. In Scotland, that limit should be at a little over 2,000 feet; "should be" because although the treeline is still there, there are no trees anywhere near it. Actually, that's not quite true – there is Creag Fhiaclach. If you were to clamber up through the Scots pines that still throng the flanks of Coire Buidhe above Loch an Eilean in the northern Cairngorms, you will eventually arrive at that melting point in landscape evolution where the corrie gives way to the plateau's almost limitless repertoire of winds. Now watch what happens. All the way up through the corrie, the pines have been growing in the normal way, and – naturally – they have thinned out and their stature has diminished towards the upper reaches. But then, and with a suddenness that startles, the trees shrink to waist height and then to knee height, and they have taken to growing horizontally. You wade among treetops. Welcome to Creag Fhiaclach, and welcome to what's left of Scotland's natural treeline. It's about 400 yards. That's it. That's all there is. But consider how many hundreds of miles of the 2,000-feet contour there must be in Scotland, then consider the astounding potential which would flow from a restored treeline. The only limit on that potential is our imagination, for once such an undertaking begins, nature's imagination would be literally limitless. "Rewilding" then, if it is to have lasting significance at all, depends on scale. Scale is everything. "Local alleviations of biotic pain" won't do. So why don't we consider restoring the treeline? All of it.

Of course, it cannot be restored in isolation, for a naturally occurring treeline must be fed and sustained from below. Essentially, what should be planted is a skeleton forest, within which nature will reassert its own priorities, its own diversity, its own densities of trees. And with our already steadily growing beaver workforce, wetland will reassert a crucial presence, crucial to the well-being of the forest, crucial to the flourishing of biodiversity. With the wolf to spearhead nature's reworking of the old order – the human order – the very land will begin to recover its health, and everything in nature will be better off and so will we. In a chapter called "The Land Ethic", Aldo Leopold had this to say on the subject of our relationship with nature:

The land ethic simply exchanges the boundaries of the community to include soils, water, plants, animals, or collectively: the land…a land ethic changes the role of Homo sapiens *from conqueror of the land community to plain member and citizen of it. It implies respect for his fellow members…*

So that's why, wandering alone among the quiet places of the land, especially somewhere like Rannoch and the hills of the Black Mount beyond, I sense an absence that I put down to wolves, and that's why the trees I hold dearest are those that remember wolves. Trees and wolves and an ice cap we have come to call Rannoch: these were the heart from which poured the lifeblood that gave birth to the Highlands we know. The heart is still there, it still beats, it still generates lifeblood, but after such a long winter has left it weary and weakened, surely a second spring is at hand for the wolf.

◉ ◉ ◉

That was how I had originally finished this chapter, but the next day, and by one of those extraordinary coincidences that occasionally grace this nature-writing life, an email arrived, an email about an extraordinary coincidence involving wolves. It came from a wildlife film-maker called Raymond Besant. I quote it with his permission:

Dear Jim,

I've recently arrived back in Orkney after filming in Tibet. I was there primarily to film the Chiru (Tibetan antelope) migration but also the wolves that follow them to prey on the newly born Chiru.

I regularly spent 14 hours a day in the hide and apart from the first few and last hours of the day very little happened! I kept myself occupied by reading and very much enjoyed your book The Last Wolf *(Birlinn, 2010) whilst there. By some fantastic coincidence I was reading the book when this lovely fellow appeared on the horizon and much to my surprise headed straight for the hide.*

It didn't stop until 5 metres from me and spent a minute sniffing the air in my direction and would have very much been able to see me through the filming windows. My heart was pumping in my chest! I was as high as a kite for an hour afterwards going from feeling very calm and privileged to utter elation.

Strangely, some of the Tibetan folk there said it must have been a "weak-minded" wolf to come so close to a human. My immediate thoughts were the opposite, that it was a confident, curious wolf checking out something new and strange in its environment.

Attached to the email were three wonderful stills of the wolf at close quarters, and one of my wolf book on his lap in front of his camera. I loved his reaction – calm, privileged, utter elation – because it demonstrates that the case for wolf reintroduction is not only the one ecological essential in a northern-hemisphere country like ours, but it also has the capacity to enrich the lives of all of us who cross the path of a wild wolf, and if we should be so fortunate we will all dine out on it for the rest of our lives. Wolves are not just good for our land and for nature; they are good for us too. Also, I was seriously chuffed that my book had been to Tibet and come within sniffing distance of a wild Tibetan wolf. I plan to dine out on *that* for the rest of my life. Thanks, Raymond.

Chapter Fifteen

The Sanctuary (2) – Loch Tulla: Desolation and Reassurance

Pinewood Hours
A half-hearted blossom
of winter-into-spring snow
time-lapsed open on juniper stems
one long pinewood hour,

and sitting in the roots of a tree
so sure of its own history
it remembers wolves,
I did nothing at all

but wait for the pinewood
to speak. A new hour began
in the same old silence
as the snow stopped falling

and spring licked its finger
to test the shift in the wind

while snared snow-flowers shrank
to dewdrops of bright white light.

That was the pinewood's message
for me, that was its gift,
that and a scrap of new knowledge:
pinewood hours take longer.

THE BLACK MOUNT IS a western outpost of Rannoch if you
approach from the east, or an eastern outpost of the historic
realm of Glenorchy and Inishail if you approach from the
west. In my mind's eye as I see it, and nature's mind's eye
as she sees it, the Sanctuary extends from Rannoch in the
north and east to Loch Tay / Loch Dochart / Loch Awe /
Loch Etive in the south and west. If you follow that progression from Rannoch, east to west, sooner or later you
will part the screen of a sudden pinewood and find Loch
Tulla there, as wild as it is beautiful, robed in trees, besotted
by mountains.

A tiny island not far from the shore harbours a clutch
of larch trees. The sea eagles that used to nest there every
year were blasted to oblivion in the 19th century. Their
near neighbours on a birch-clad island at the western edge
of the Moor met with the same fate. But this was always
eagle country, and it did not stop being eagle country just
because its eagles were (what's the word?)…euthanised.
Besides, if I think I can sense the land's mournful remembrance of wolves, surely this century's generations of wandering young sea eagles raised on both coasts can surely
sense (and mourn) the absence of their ancestors as they

drift this way with increasing frequency, for what they see is a land fit for sea eagles and ready to be re-colonised. They also see that golden eagles still live here, for they somehow clung on despite the slaughter. They were always warier of man and his guns, always presented more elusive targets, and today the presence of golden eagles is a good omen for the sea eagles. The old island nesters of Loch Tulla and Rannoch Moor could both see the comings and goings of the golden eagles that nested at a certain crag not far away, and when the new sea eagle generations reoccupy the same sites they, too, will see the comings and goings of the new golden eagle generations at that same certain crag.

As for the eagles of the crag and the air and the water, so for the wolves of the mountain, moor and forest. This was wolf country, too, and for the same reasons: the wildness, the space, the food supply, and the more or less limitless opportunities to reach out to new territories in every direction and so extend their presence, to lay claim to the Sanctuary. And those other great manipulators of wild country in nature's cause, Scotland's reintroduced beavers, are already nibbling around its edges. Serious inroads are a matter of time. The world-famous reintroduction of wolves into Yellowstone in the northern United States has demonstrated that wolves and beavers work in tandem to transform biodiversity. This land, this Sanctuary, is where nature can begin again to effect cures.

I stepped this way one early May day of that backward spring, and I went first in search of its ospreys. In the story of the osprey's return to Scotland of its own volition, which began in the 1950s, the ospreys of Loch Tulla were comparative latecomers. If you know where to look, there is a sightline through trees to an osprey eyrie. Further south and east they had bucked that season's trend of migrating birds by turning up right on time, but it became obvious at first glance through the pinewood to where the nest tree stands on the loch shore that here they had not turned up at all. The nest, what was left of it from the previous year after a winter's mauling by wet gales and snow winds, slumped at an angle that suggested one more healthy gust could tip the whole edifice onto the beach. It was a forlorn ruin with a tuft of grass growing in what had once been redoubtable timbered walls. That grass was the nest's only sign of life. It was possible, of course, that the ospreys had abandoned the site and moved further up the shore or across the loch, but that day and a return visit ten days later would reveal no glimpse of them, no sign of a new nest.

It was a troubling omen, for I had shown to my own satisfaction that waters where ospreys thrived were also likely to suit wandering sea eagles looking for territories as their recovery spreads across the face of the land. Seton Gordon wrote very specifically about both sea eagles and golden eagles right here, and as evidence grows of sea eagles returning to historic sites, sites where they once thrived more than a century ago, I have high hopes for this one. It seems perfect in every way. But as I write this, I am no nearer to knowing why the osprey nest was not rebuilt. I am consoled, however, by personal observation of young sea eagles moving in both directions between Scotland's east and west coasts, and more often than not their line of travel is along a swathe of latitude roughly between the Tay estuary and Mull, so Loch Tulla is perfectly placed and it is surely nothing more than a question of time.

Meanwhile, early May around Loch Tulla was not looking like itself. The larches wore only the mistiest, palest of green tinges; a square yard of wood anemones at the foot of a pine tree was the only flower in sight; the hill grasses simply looked defeated; the mountain overlord of Stob Gobhar across the loch was still scarfed and banded with deep snow, and a new powdery fleece had gathered around the summit in the night. I headed for the shore. There are often signs of otters along the banks where a quiet river uncoils from the west, so I would start there.

It was the very last few inches above the shore where spring suddenly stirred: a pair of common sandpipers I had simply not seen sped out from a hidden patch of muddy ooze, inches above the water, inches above their own reflections; these are birds that deal in inches. They piped in high-pitched triplets and quadruplets, curved in a surface-skimming arc to a new shoreline ooze where they landed, flared their wings inches-high above their heads, and fell silent, tails working like silent metronomes. It was a start.

Then the cuckoos.

Out of profound silence two birds began a precisely executed call-and-response routine, so precise that the second bird sounded like a perfect echo of the first, so perfect that I wondered if it *was* an echo. But then the second bird moved closer to me and the first one stayed put, and still they called without a break, on and on and on, cuckoos on a loop. What finally stopped them was the intervention of a third bird, a third call, a wrecking ball for the rhythmically perfect call-and-response of the first two birds. They stopped calling at once. But now the third bird went on alone, and to add to the chaos it had already caused, its call was "wrong". The standard, production-line cuckoo call is a minor third. This one was a major fifth. But the high note was imperfectly formed, for it emerged as something like a corvid squawk, as if the bird was determined to hit the note right in the middle but lacked the range and had to reach for it, and never quite got there.

It is still the only time I have heard such a thing, but once, up in the golden eagle glen I know best among the mountains of Balquhidder, a jazz cuckoo turned up and lingered for weeks. Instead of the two-note, minor third production-line call, it had a third note. But the three notes were not equally spaced. The third note had a slight delay built in, so that the rhythm it produced was that of a jazz waltz. As a part-time jazz guitar player myself, and as one who has composed a jazz waltz (mysteriously titled "A Wee Jazz Waltz"), this particular cuckoo was a particular delight. On days when the eagles failed to show, it would brighten the mood with its jazzy voice. What I wonder is whether the distinctive voice passes down the generations or whether (as I suspect) it dies with the singer, in much the same way that Frank Sinatra Junior ain't no Frank Sinatra.

Eventually, all three Loch Tulla cuckoos fell silent. I wondered how they would fare in such a spring. Because the insects were drastically depleted by the overlong winter, so were the insect-eating birds – like the pipits that cuckoos rely on to bring up their monstrous chicks, with their perverse instincts and built-in egg-eviction scoops on their backs, which clear pipit eggs from pipit nests. Sometimes nature works in mysterious ways. I don't believe for a moment that there is such a thing as "the balance of nature", and whenever I hear anyone extolling the virtue of such a phantom concept I offer up the same response: it's cuckoo.

◉ ◉ ◉

"*Tiu-tiu-tiu, tiu-tiu-tiu…*"

Ah, that new voice is no cuckoo. And in this corner of the Highlands it is as rare an utterance in the conversation of spring as the cuckoo is ubiquitous. I had expected to find ospreys and found none; in my wildest dreams I did not expect greenshanks, and I found two. They came towards me as low over the water as sandpipers, wingtips almost touching, then split to two different scraps of shoreline fifty yards apart. As each bird landed, one three or four seconds after the other, it also displayed the sandpiper habit, the habit of many waders: it raised elegantly angled and pointed wings high over its back, and these were silvery white in the still reluctant sunlight.

Greenshanks dress to blend in. They wear the shades of summer peat bogs and bog cotton; or, on the wrong kind of sluggishly cold May day, the shades of peat bogs and snow. The shades in question are dark grey and light grey and white and green and greyish-brown and greyish-white and greyish-green. On the nest, which is typically a half-hearted scrape in the surface matter of Sutherland and Caithness or Lewis and Harris (so cool, wet, rocky, peaty, watery, windy), the brooding bird sits in the lee of a small rock or a piece of ancient bogwood, and the chances are you won't see it at ten paces, which is why the greenshank dresses to blend in.

But greenshanks are also the most contradictory creatures in the Northlands. They are waders that don't flock. And while they may wear the shades of peat bogs and bog cotton in order to be more or less invisible, they will insist on opening their mouths and giving the game away. Even the bog-standard contact call of the greenshank (never mind their high-up and handsome pre-nuptial, pair-bonding carol) can water the eyes and gladden the heart of ornithologists. It says "*cheep*", but it does so with a silence-stabbing potency and a hint of impending pibroch that are the stuff of birder eulogies. If you are fortunate enough or skilful enough (usually both qualities are required) to be present within sight of a nest when the birds change over on the eggs, the approaching bird announces its arrival, and then its presence, with an outburst of extrovert cheeping that can last for ten or fifteen minutes, and if you don't know what you are listening to by that time, then you need to get out more.

And to further undermine its low-key presence in solitary, wild places, when it flies away from you it flashes a white blaze of feathers in the sunlight that tightens your throat and confirms your diagnosis, and you mutter under your breath to no one at all: "Ah, greenshank!"

My limited experience of them has tended to occur in memorable places. Tentsmuir beach in the north-east corner of Fife is one. At low tide and in late winter, the wind rummaging among the dunes with a hint of Svalbard on its breath, the place has a primitive Arctic-edge feel to it, a trait emphasised by drifts of old snow lingering here and there on the lee side of the dunes. I had stopped to photograph the eerie presence of snow on sand in the lowering light of late afternoon, glad of the respite from the wind for a few moments. I cooried into as wind-free a corner as such a day in such a landscape can contrive, and happed my hands about a flask-cup of coffee with more than a hint of something peaty and Hebridean in it to thwart the chill and to mask the corrupting consequences inflicted on the taste of what was once good coffee by the malevolent influence of two hours inside a flask. I let my eyes wander while I was trying to unravel a possible scientific explanation for the coffee thing, and a flicker of movement lodged at the extreme left edge of the dunes.

Long, shallow pools of water often lie in among the last of the dunes before they yield to the wide sprawl of the open beach, and a few minutes before I took my photographs I had to divert to circumnavigate one such pool. In fact, so low and flat was the light that I had thought at first that its white disc was one more patch of old snow. From where I sat I could only see one small arc of the water, but it was there that something had caught my eye, and now it wasn't there. It could have been anything, but instinct plays a significant role in my relationship with the natural world, and instinct now prodded me in the ribs, so I swallowed the last of the Talisker-infused coffee, stowed the flask in my small pack, and slipped away in a wide detour to come on the pool from the shelter of its nearest dunes. The nearest waders were a hundred yards away, and beyond that they had gathered in improbable numbers that looked as if they stretched most of the way to Denmark. But in the pool that came into view through a flimsy screen of wispy, wind-tugged grasses on the top of the dunes, there was a solitary bird. Its long legs were pale yellowish green, and its rump, which was raised towards me, was a blaze of white, and I muttered under my breath to no one at all: "Ah, greenshank!"

I know enough about looking at stained glass (because I was taught by one who knew a great deal about stained glass) to choose the dull days, for it is then and only then that the glass catches fire. Likewise, the best conditions in which to view a "drab" bird like a greenshank are not to be found in days of blue-skied sunlight, but on days like that one of thick, fat clouds welded together to smother every square mile of open sky, when the sand dunes are as dulled as cold tea, and late winter is drawing its blinds down to meet the advent of dusk. Then, that solitary greenshank, that haunter of lonely shores and wild, unpeopled tracts of the north of the land, is suddenly and mysteriously aglow, a creature of some finery. Every dark-greyish feather on its back and upper wings is edged in white, which imparts to the bird an air of stained-glass delicacy quite at odds with the kind of life it leads. The white of its underparts rises up towards its neck and throat where it is striped with black, stripes that break up into orderly lines of tiny patches. The bird zipped around the shallowest stretches of the pool, sifting it with lowered and ever so slightly upturned bill. It moved in mazy, unpredictable courses, and in the flat white light that bounced back from the water and the snow, this single greenshank glowed like a fragment of one of the windows of Chartres.

These new arrivals at Loch Tulla were no winter refugees, but a pair that showed every indication of nesting. And even though they were a long way south of what one might think of as their breeding heartlands, there is something about this loch and its immediate surroundings that reeks of wildness, that beckons to what a nature writer is apt to think of as symbols of wildness, and the greenshank is certainly one of those. If you come on the place having crossed Rannoch Moor, or (as is almost invariably the case in my travels) from the southmost edge of the Highlands through Balquhidder and Glen Dochart, or even from the north – from, say, Glencoe or the Ben Nevis range – there is a sudden sense of an oasis, of a harder-edged, thicker-skinned landscape that seems to belong directly to none of its surrounding lands and watersheets.

The pinewoods have something to do with that. Native woodland invariably adds a wild edge to a mountain landscape that is missing where the trees have been wiped out by the dead hand of deer forest and grouse moor management. Rannoch Moor was a woodland, the souvenirs are everywhere buried in its peat, bits of trees that – if you could ask them – would tell you all you ever needed to know about wolves in the Sanctuary. Where the woodland is still present, to one extent or another, nature tries harder, and those wilderness-thirled species that find it so hard to find a home in Scotland are drawn to the Loch Tullas of the land. Witness the frequent encounters with pine marten droppings in the woods, along the tracks and even on the road. And now, having been troubled by the desolation of the osprey nest, I was reassured to some extent by the discovery of a pair of greenshanks.

One of the birds had perched by the mouth of the river and began walking the shore of the loch towards me. The second bird called twice then flew across the bay to land right beside the other. They came on together, stabbing the mud, sifting the water's edge, reaching into the nearest tussocks. Everything about their behaviour suggested they were newly arrived, that they had yet to contemplate a nest site (the male doesn't so much build a nest as scrape a choice of three or four token depressions in the land; the female chooses one of them, apparently on the basis of it being somewhere to sit and better than nothing, but not much). For the moment, they were flying and feeding companionably. Abruptly, they turned back along the shore then flew into the mouth of the river and vanished among its high-banked bends. I set off slowly and quietly in their wake, with the secondary intention to check for otters. A leisurely exploration of the mouth of the river revealed some old otter spraints, no new tracks, an astonished pair of greylag geese that burst up from the quiet river in an explosion of big wings and far too many decibels for the good of my intentions, for their explosive flight away across the loch had alerted my presence to every other tribe of nature within half a mile. I quietly withdrew and promised myself to return in around two weeks to check on the greenshanks.

Late in the afternoon, driving out along the single-track road to Bridge of Orchy and the road east, there was a last glimpse through the pinewoods to the loch. I was driving slowly, the road was empty, and I was happily rewinding the day through my mind while keeping one eye on the road and the other on the land as it drifted by. There was a small flash on a last corner of the loch but vivid enough to be eye-catching. It was a fragmentary thing, a tiny flaring of bright white, but a thing that struck the corner of an eye in a way that roused that instinct which I trust so utterly. I pulled off the road to investigate but already I thought I knew what I had just seen, for it was another of my native land's ambassadors for wildness, and I had glimpsed it occasionally on Loch Tulla over the years. There it was again, in the glasses this time, the same heaving movement and just a few yards out from the shore. It was a movement like a boat rolling, baring the hull all the way to the keel, then righting itself, the whiteness diminishing then vanishing as the motion restored to its proper place a superstructure immaculately patterned in black and white and marbled grey. Behold the black-throated diver preening as it swims.

Then I realised there was a second bird just ahead of the first, swimming slowly towards a small and thickly wooded island just offshore, the perfect setting for a black-throat nest. They would like the cover, and as long as there was a short and easy route from the water to a flat nesting area that involved no more than two or three yards of walking, it looked ideal. The walking distance would be crucial, however, for although the black-throat is a supreme swimmer both on the surface and underwater, and a powerful flier, its legs are set so far back in its body that it walks with all the aplomb of a fencepost. That second bird, which I took to be the female, vanished behind the island and did not reappear, but its mate remained in full view, and going nowhere fast. Crouching by a broken old pine trunk, and with my head already full of the day's events and an atmosphere that had led me to pursue a train of thought about the particular species of wildness I always sensed when I came here, I suddenly had a strange moment of déjà vu.

It involved a different diver on a different loch, and for that matter a different pair of binoculars, for it was more than thirty years ago. I had gone to visit an old friend, the late Mike Tomkies, at the isolated Loch Shiel cottage he called Wildernesse. Mike was the most uncompromising nature writer I ever met, and in the early days of a friendship that sadly did not go the distance, he was incredibly generous and encouraging to me. He had done some groundbreaking work on his local black-throats, and he was raving quietly to me about them and what they meant to him while we watched a pair joust with the waves that could suddenly tear across that loch with the breath of an Atlantic storm about them. I can do the moment no better service than to quote a passage about them in his best book, *A Last Wild Place* (Cape, 1984):

> *For me they embody as no other creature does the wild spirit of the loch…when flying high on their short but powerful wings they look like arrows of twanging steel…Their magnificent summer plumage, with its rich blend of slate greys, blue greys, purples, blacks, creamy underparts and sooty throat patches must be the sleekest of all water birds', and their array of intricate white neck stripes and snowy wing bars, if you're lucky enough to get close, dazzle the human eye…*

He also described it in purposeful swimming mode as "steaming along like a small barge".

On Loch Tulla that cold day at the beginning of May 2018, the "barge" in my binoculars was idling, and there were the stripes and there were the dazzling, snowy wing bars. I was as sure as I could be that these two divers were not yet nesting, that they somehow epitomised the particular aura of wildness that I always sensed on this high, mountain-and-pinewood loch, that they functioned almost as an ecological barometer of the place, and that – like the greenshanks – they were waiting for that backward spring to catch up, and that if it did not catch up soon, they might simply not nest at all. My mind drifted from there to those other rarities of nature in the Scottish Highlands that inhabited the place – the greenshanks, the wandering sea eagles, the native golden eagles, the otters, the pine martens and the ospreys that had so mysteriously failed to appear; and how I would love to know if somewhere in the wildest fastnesses of the loch's mountains and pinewoods there might not also be a surviving pocket of wildcats. If such an isolated survival can happen anywhere, it can happen here. Two weeks later, the very middle of May, I was back, and I was looking for all of the above, but more than anything else, seeking reassurance about the greenshanks and the divers.

The day was warm, the first really warm day of the year, as long as you kept something solid between yourself and an edgy east wind that had barely paused for breath since it ushered in the "Beast from the East" at the beginning of March. There were new pine marten droppings right at my feet as soon as I stepped in among the trees. I headed straight for the mouth of the river with the intention of working back from there as far along the loch shore as inclination led me. I noticed as I walked that there was no hint of new grass anywhere, that the mountains were still ridged and heftily corniced with snow. The river revealed nothing at all, absolutely nothing: no living mite, no tracks in the mud, no otter spraint, no greenshank call, no sandpiper piping. It was as if the season had gone backwards, despite all the new warmth and the sunlight. I turned and walked the loch shore, a narrow strip of bog and tussock between the water's edge and the birch and pinewoods that climb steeply there. The big larches out on the headland looked marginally greener, and as I scanned the woodland I admired alder, willow, rowan, holly in small cliques, but only the holly was green.

I reached the osprey tree in a mood that was tumbling towards something uncharacteristically morose. Then a greenshank flew from the very edge of the shore and only seven or eight feet in front of me. I had not seen it; it had been sitting tight until it lost its nerve. I thought for a moment I might have disturbed a nesting bird, but it flew in complete silence, and a thorough search revealed nothing at all. Its footprints were everywhere in the mud, however, so it was reasonable to conclude that it had simply been feeding. It flew about 200 yards to a spit of land that jutted into the loch, and there it vanished, and I saw no sign of it thereafter, no sign of its mate. The osprey nest appeared to have tilted a little more. I pored over the loch, found some wigeon and teal by the far shore, the two greylags, and nothing else at all, certainly nothing that looked like a black-throated diver.

I turned the glasses on the mountains. I know one golden eagle eyrie just a watershed away, and sometimes on previous visits I have seen the birds cross it in both directions from the lochside. I watched for over an hour – the sky, the ridges, and I could hear Mike's words in my ears, for he was a great student of golden eagles: "you must learn to scan the middle distance". The snow on a long stretch of ridge had melted from a series of buttresses just below the ridge, creating the impression of a system of arches, a sort of viaduct of solid snow, an extraordinary illusion that blazed in the sunlight. But no eagle shadow crossed the viaduct. I had come that day with two ideas in mind: one, to catch up on the greenshanks and the divers; the other, to check on the progress of spring. Now I decided to move to a high wedge of pinewood so that I could sit and watch the island where I had seen the divers. I found a broad pine trunk with a flat ledge among the roots and a cushion of pine needles, out of the wind and in the sun. My job does not get much better than this.

I had a long, lingering look at the island in the glasses. The two sides of it that I could see were thickly wooded right to the very edge, a few old and tall birches, the rest of it crammed with alder. I could make out nothing that confirmed either the presence of a diver nest or the complete absence. It was always possible that on the first visit the birds were simply prospecting, exploring various possibilities before they felt ready to nest, or perhaps before they felt that the laggard spring was ready for them.

I settled in for a lengthy stay, ate lunch about 3p.m. (it's the sort of thing that happens on such occasions, suddenly I think: "Oh, food. I should eat.").

There was a wide view over the bay beyond the island towards the south-east corner of the loch and to the mountains beyond. As the afternoon advanced, the sun warmed and the wind that roughened the surface all day fell away and the loch grew calm. It also changed colour over an hour or so from serried greys to almost colourless white to sky blue and then to an ever-deepening blue as the sun shrugged off the last of the clouds. Every few minutes I quartered the loch and island with the binoculars. In between these searches I watched the trees, the sky, or just let my eyes wander, scanning the middle distance. Sometimes, when you just sit, effectively become pine tree or mountainside, or in this case, both, nature sends out an emissary to check you out. Not today. Nothing moved. I opened a notebook, laid a pen across its open pages. Sometimes you have your most original thoughts when you are out there and nature drops something priceless into the lap of your stillness. An hour later, the pen still lay where I had put it, the white paper unbesmirched.

I looked at my watch finally. It had been five hours in all, three of them sitting right here. These are the days when you learn a little about the pace at which nature sometimes moves; these are the days when you learn a little more about yourself in nature's company – how you respond when she makes unusual demands of you. Then somewhere between 300 and 400 yards out from the island, a moment of bright white amid the deep blue. A breaking wave stirred up by a stray gust? Or…even in the binoculars I was unconvinced, for it vanished as soon as it brightened. Then it stood on its tail, opened and flexed short wings and the bright white flared again. Then I saw it dive, and it was grey and black as it dived. And that was it.

One glimpse of a greenshank, one of a diver. Nothing more, but oddly, it was reassurance enough. I have never needed to know where the nests are, I was never that kind of nature writer. Nests are places where people should not be. Nests are what the wildlife needs to get by, to maintain their toehold on a scrap of land. We should be no part of that. What matters to me is that they are here. That they take their place in the biodiversity of this extraordinarily beautiful loch. Of all places, I learned that lesson most vividly right here. That and the certain truth that pinewood hours take longer.

Mike Tomkies' last word on the black-throated divers was this:

Here, like the eagle or the wildcat, is a creature beyond man, rare because it is beautiful, rare because its needs are sensitive, rare because it requires solitude and has for centuries survived in true wildness. It is the true spirit of the magical loch... The loch will outlive me and possibly all human life and all I can do is seek a little longer to know its heart.

Amen to that, Mike.

Chapter Sixteen

The Properties of Mercury

SUDDENLY I REMEMBERED O-LEVEL chemistry and mercury, the metal that seemed to exist in a scientific no man's land between solid and liquid, chemical symbol Hg. It could slink across a slightly tilted surface like thick oil, though it declined to drip. But you could cut it into pieces that slunk independently of each other and bounced and rolled off each other rather than coalesced into a single slink again, like thick oil would have done. That, at least, is what I remember, and while it may be a flawed memory (chemistry was the only O-level I failed), what I was watching reminded me of it. I also remembered that the effect had been slightly unsettling, as was the thing I was watching.

That memory then, had hibernated undisturbed for more than fifty years, so unnecessary to my subsequent life has O-level chemistry (failed) and a rudimentary knowledge of the properties of mercury proved. What awakened it from its long slumber was the small spectacle of an out-of-sorts movement that shone and moved – mercurially – along a ploughed furrow, the thing rendered unearthly by low morning sun on a ground mist that clung to the last south-facing slope of the Highland Edge. By the time I had wiped and focussed the binoculars, the thing – the procession of mercurial particles, as I saw it – was bearing down on me at speed, for I was passing the end of the relevant furrow when it caught my eye. What I had seen was a family of stoats on the march. No, wait. "On the march" won't do, for stoats don't march any more than mercury does. And a posse of seven stoats hemmed in by the walls of the furrow and moving straight towards me in not-quite-single-file looked like a snake running. Or like spilled mercury slinking across the scarred surface of a 1960s science lab bench.

Nothing much fazes a dog stoat, although he has good historical reasons to turn aside from a man at the far end of his chosen furrow. But his eyesight is not the sharpest tool in his armoury, so the stoat-snake with the dog for a head ran on down the furrow. His scenting powers are astounding but he wasn't getting mine thanks to the in-my-face breeze. His hearing is as good, but I wasn't making noise. He got to ten yards away and stopped. The rest of the snake stopped behind him, somewhat chaotically, as if someone had tried to apply brakes to moving mercury.

The dog stoat stood up on his hind legs, which, as with everything from a weasel to an otter to a badger to a wolverine to a black bear to a grizzly bear to a polar bear, is a sign that he is curious. He wants a clearer view, a better scent, more information. Then, in a routine that would not have disgraced a Disney animation, the bitch stoat stood and five miniature stoat kits stood, each one trying to crane forward round the head of the one in front. Again, the effect was vaguely unsettling. The dog stoat can stretch to about ten inches tall when he stands, the bitch maybe two inches less, and the kits to about halfway. And I am half an inch short of six feet tall. So my binoculars were more or less full of tiny pairs of black eyes, bobbing and swaying on rigidly planted feet, and I was trying to imagine what I looked like in those tiny eyes while also trying not to laugh out loud.

I wondered what difference it would make if I got down to something closer to their level, so I sat. The answer to my wondering was, no difference at all, except that I had a vague idea of what Gulliver must have felt like among the Lilliputians. So I tried making kissing noises, a ruse well known among keepers and wildlife photographers and just plain wildlife watchers like me who, for very different motives, like to bring stoats within range. That worked, and then it didn't, for the dog stoat ran forward two yards then stopped again, with one forefoot poised on the up-slope of the wall of the furrow. Behind him, the bitch and the kits lined up, each one slightly higher up the wall of the furrow than the one in front, arranged like half a skein of geese, and I wondered if that was an involuntary fluke or whether it was a strategy.

At that point, a tractor barged up the farm track beyond the field edge and galvanised the stoats. They flowed up and over the furrow and into the next, then over that and into the next, then the next, the next, the next, so close together as they flowed over the contours that it was almost like watching a single otter rather than seven stoats.

Then they flowed into one more furrow and did not reappear, and I guessed they were running along it back the way they had come. And they told me in the science lab that mercury couldn't flow uphill. No wonder I failed chemistry.

⊙ ⊙ ⊙

There was a time in my life when I saw a stoat more often than I saw other people. A dog stoat was a reliable seven-days-a-week, four-seasons-a-year visitor to a dyke in the garden of the Glen Dochart cottage where I lived for a few years. In the course of a year his coat was every known patchwork variation of chestnut and snow, the black tip on his tail the only constant, his tribal badge. The dyke was shoulder-high and drystane. It was about twenty feet long and petered out into a climbing bank between the garden and the road beyond. It separated (with a field-sized gate, and for no good reason that I ever fathomed) the front garden from the back, and it was as well made as it was apparently pointless. I used to see it every time I lifted my head from the writing page and looked out of the window, and there were times when its small wildlife dramas used to drag my head away from the writing page, demanding to be looked at.

I had a field mouse problem at the time. The cottage had been uninhabited for a few months before I moved in; or, more accurately, it had been uninhabited by people. But field mice inhabited every room, including a nest in the living room sofa. I like to think that I am the most tolerant of hosts and neighbours when it comes to living with nature, but a destructive invasion of my living space tends to make me less tolerant, less neighbourly. After several non-lethal attempts at a solution failed, I resorted to lethal ones. I caught thirty mice in two weeks, after which the problem declined to manageable proportions. I began to put the corpses out on a flat rock in the back garden. The first one lay there for a day and a night before the neighbourhood sparrowhawk saw it. After that, they were rarely on the rock for more than an hour, for an inspection of the rock became an integral part of the hawk's hunting regime. The hawk thought me particularly tolerant, particularly neighbourly.

The dyke had a bird feeder that would have delighted Heath Robinson. I had found the discarded head of an old garden fork nearby, and while I mulled over the possibilities of what use I might make of it, if any, it occurred to me that if I jammed its prongs between stones in the top of the dyke, I could hang bird feeders from its other end. Field mice lived inside the wall, and established themselves as regular if unskilled tightrope walkers, gingering out along the fork for the prize of a peanut and teetering back to the sanctuary of the dyke. Accidents were frequent – they fell off. Invariably they hit the ground running and were soon trying again. It was when a peanut turned up in my bed that I realised I was in for the long haul in my campaign to keep my house reasonably mouse-free. I don't eat peanuts, in bed or anywhere else, and the attendant mouse droppings were as good as DNA in securing a conviction. So I stepped up my pursuit of the criminals.

I was aided in my campaign by the dog stoat. I saw him one day on the crest of the dyke, standing on his hind legs and arching his whole body to peer first down one side of the dyke then down the other, while his hind feet remained planted. Then a field mouse face appeared by the fork shaft. The stoat dropped to all fours and froze. The mouse advanced. The stoat leaped. For an airborne split second, the mouse's fate hung unresolved as it advanced a few inches along the fork shaft. But field mice are wary critters, and some sixth sense told this one that all was not well within the force field of the dyke. The mouse reversed, for all the world like one of those figures that emerge and retreat from old clocks to pronounce a change in the weather. So it was just inside the wall when the dive-bombing stoat hit the fork with its front feet, failed to hold on, then spun away in a free-falling curve to crash-land in some disorder in the grass. The empty fork quivered.

Then the stoat shot forward into a ground-level hole in the dyke and all hell broke out. As he entered the bottom of the dyke, a mouse appeared on the top, and another by the fork. The stoat reappeared halfway up, flowed up over an overhang and into another gap. A blackbird and a wren both appeared at opposite ends of the top of the dyke and began screaming abuse. The topmost mouse vanished. The other one was inexplicably trying to dislodge a peanut from the feeder. Not now, I told it, not now; then I decided that I was on the stoat's side, so why not now?

The other mouse (I guessed there were only two because I never saw more than two on the dyke at the same time, which is hardly convincing evidence) and the stoat must have covered every square inch of the inner and outer surfaces of the dyke with breath-catching fluency and barely a pause or a false step. Each time the stoat broke cover the blackbird and the wren screeched, but the stoat flowed heedlessly between them as if they were clumps of moss. Nor did it pay the slightest attention to the mouse at the nut feeder which had apparently decided that stillness was a virtue in these circumstances, that outside the dyke was safer than inside, and that if it had to be outside and still then it may as well be outside and still by the feeders.

Finally, the stoat emerged backwards from the ground-level hole where it had first entered. The other mouse was dead in its jaws. Blackbird reinforcements arrived, and as the stoat vanished round the gatepost and into the wilder tracts of the back garden, I could follow its unseen progress by the diminishing volleys of shrill abuse.

Such small life-and-death duels of nature are acted out by the million everyday and all across the wild world. Yet if you witness even one of them from beginning to end you are one of the fortunate few, for to bear that witness is to understand nature a little better. And if you do witness such a duel, the chances are you were sitting still at the time. It made me wonder what I must have missed every time I went into the kitchen to make a cup of tea.

◉ ◉ ◉

The stoat is an acrobat and a clown and has an appetite for play at least as hearty as the otter's. I once watched one play with an acorn for ten unbroken minutes, lying on its back and juggling it with its four paws, "dribbling" it along a path, the acorn never more than a couple of inches from the end of its nose as it ran. After twenty yards of path, the stoat U-turned and dribbled the acorn back to where it had begun, then started tossing it into the air and leaping to catch it, often completing the catch with a victory roll on the ground. Then it got bored, dropped the acorn, and vanished.

Play in such an animal is a many-sided virtue: it bonds a family, it hones agility and reflex and co-ordination, it's fun, and it loads the odds in the stoat's favour when he goes to work as a predator. Whenever people talk about stoats, stories always crop up of how they mesmerise birds, rabbits and even hares by performing an apparently limitless repertoire of party tricks; then the instant transformation from party animal to predator, the lethal bite. And some of the stories are even true. I have heard experienced naturalists insist that if you disturb a stoat when he has "hypnotised" a rabbit with his deadly dance, you can then pick up the afflicted rabbit until it snaps out of the spell. I'm not clear why you would want to – presumably the rabbit will snap out of it without being picked up, and, in any case, why traumatise it twice? Sometimes the stories get preposterously tall – a stoat killing a cat with a single bite is one I've heard – but such stories are part of the mystique of stoats, part of their endless fascination, whether you want to marvel at them or string them up on a gibbet.

One snowy April, when attempts by an organised watch to keep an eye on a vulnerable golden eagle eyrie were hampered by a thick wedge of snow on the eyrie ledge that completely obscured the sitting bird, I saw the male eagle homing in on the crag with what appeared to be a white rag in its talons. Just as it raised its wings high before alighting on the eyrie, I saw through the binoculars what I'm fairly sure was the lifeless form of a stoat in ermine. It must have known very little about its last moments. Clearly, when the eagle swooped, no one had bothered to tell it that you can't pick up mercury.

Chapter Seventeen

Renaissance

The Hopeful Oak

Salome she isn't
(she dances without veils)

she's quite dead
but still dances

naked and skeletally elegant
trunk straight (dead straight)

three limbs six branches
upraised broken-ended

yet lovingly tended
flimsily screened

scantily greened
by three birch saplings

Salome she isn't
but even naked and dead

she reaches hopefully
for the stars

GLEN FINGLAS HAS BECOME central to my writing life. In the landscape I think of as my nature writer's territory, it lies on the first upslope of the Highland mountains as you enter them from the south. Behind you lies Loch Venachar, whence the local ospreys will fly across the last of the Lowland hills to their south, two miles to a low-lying, shallow and conspicuously well-stocked loch, in search of easy pickings among the anglers in their boats. Then, of course, they must gain a thousand feet of height again, this time carrying two pounds or so of rainbow trout, re-cross the hills, then glide back down to the nest tree. All of this is a much more fruitful technique than peering into the unguessable depths of the loch at the base of their nesting tree, the tree where a sitting bird facing north looks out across the loch to a mountainside that is in the throes of changing colour, of transforming the fortunes of everything that lives there; it is, truly, a landscape in the throes of renaissance.

I walk here regularly in every season and have done over the last dozen years, often following the same paths and the same routes through the higher pathless tracts, because the walking offers the opportunity to watch the renaissance of a landscape in real time. This is how you bring a lost landscape back to life. It may not change visibly from day to day (other than those endless variables wrought by conspiracies of light and weather and season), but if you continue returning over the months and years you develop a sense of nature at work in a primitive, imaginative and benevolent way. Nature's secrets are uncovered by reworking the familiar, by trying to write what I think of as the portraiture of place, and by going back again and again over years, decades, understanding it better each time. It's Cézanne again, painting that single, singular mountain. Or it's Norman MacCaig again:

So then I'll woo the mountain
Till I know the meaning of the meaning, no less…

And when you tread an unfamiliar landscape for the first time, you take the familiar with you, in your head, in your fingertips (because you have touched the trees and the old rocks, and because your pen hand has written them down) and in your heart because you have learned to love that confiding familiarity; and when you look at any unfamiliar land through the prism of that familiarity with nature and how it works, you find that – of course – the unfamiliar land works in very much the same way.

I had stepped aside from the path up through the oak-wood because the slant of ice-cold spring sunlight had illuminated a tree in a certain way so that its shape had caught my eye, and I thought I might photograph it or draw it or simply sit beside it for a while and try to write it down. It was fostering a rowan in its fork; where the trunk divided into three mighty limbs, a bowl had formed and a hundred years or so of bits and pieces of the parent tree had accumulated there, broken down into a kind of compost fuelled by a hundred years of rain water and, I suppose, the occasional accumulation of snow. Somewhere along the line a bird or a squirrel or a pine marten had disposed of an undigested rowan seed there, and it took root. I sat and scribbled, and one way or another I looked into my idea of the mind of the oakwood and tried to fathom out its sense of itself. Put like that, it sounds faintly preposterous, a fool's errand at best, yet such approaches to nature in pursuit of a kind of intimacy have been at the heart of nature writing for 200 years. Here is Henry David Thoreau writing in *Walden*:

> *One old man, who has been a close observer of Nature, and seems as thoroughly wise in regard to all her operations as if she had been put upon the stocks when he was a boy and he had helped to lay her keel…*

The first volume in this tetralogy of the seasons, *The Nature of Autumn*, opened with the arrival of autumn geese above my boyhood home when I was a four-year-old, and what was that but nature "put upon the stocks", and had I not been helping to lay the keel ever since? And what was this with the oakwood but another plank in the keel?

And what if I return to Norman MacCaig for further validation, that same poem – "Landscape and I" – from which I quoted earlier? It begins:

> *Landscape and I get on together well.*
> *Though I'm the talkative one, still he can tell*
> *His symptoms of being to me, the way a shell*
> *Murmurs of oceans.*

And later:

> *This means, of course, Schiehallion in my mind*
> *Is more than mountain. In it he leaves behind*

An idea, like a hind
Couched in a corrie.

Of course, the endeavour has its limitations. Perhaps the landscape can only "...tell his symptoms of being to me..." but the point of the endeavour at all is twofold: to enhance the relationship between the writer and his subject, and to foster a climate in which the eternal possibility exists of the miraculous, the unpredictable moment of revelation. The American Barry Lopez wrote in *Arctic Dreams* (Scribner, 1986):

> *Whatever evaluation we finally make of a stretch of land, no matter how profound or accurate, we will find it inadequate. The land retains an identity of its own, still deeper and more subtle than we can know. Our obligation toward it then becomes simple: to approach with an uncalculating mind, with an attitude of regard. To try to sense the range and variety of its expression – its weather and colour and animals. To intend from the beginning to preserve some of the mystery within it as a kind of wisdom to be experienced, not questioned. And to be alert for its openings, for that moment when something sacred reveals itself within the mundane, and you know the land knows you are there.*

Fat trunks rippled with inches-deep moss, fooled the eye with too-early-in-the-spring foliage of ferns, not oak leaves. An ambience of age pervades the innards of a mature oak-wood, assisted by stillness. I could hear wind, but it was high up and elsewhere, and the woodland floor was breathless.

But a healthy woodland is home to trees of many ages, and this one, rescued from incipient oblivion by Woodland Trust Scotland, is evolving beautifully now that the forestry industry's regime of ruthless monoculture has been banished and replaced by nature at its most vigorously opportunistic. I followed the edge of a kind of terrace where self-sown birch saplings crowded into the space vacated by felled spruce plantations from the bad old days. In the damp and eager hillsides of the Highland Edge, birch trees are the most enthusiastic of colonists, and everywhere in their wake come the willows, the alders, the rowans, the hollies, the occasional splash-and-tremble of aspens. Beyond this incursion, the oaks resumed, and I met one leviathan tree that had insisted on – and now defended – a vast tract of land and airspace. Beneath it there flourished another level terrace of short, emerald grass, and that moment of that March afternoon of the backward spring, something suddenly yielded and sunlight flooded the terrace. That was the oak's gift to me. I sat, spread out lunch, notebook, sketchbook, camera; I basked in nature's company, "…and the land knows you are there…"

The nearest oak tree to the leviathan was about twenty yards away. It was a runt, impossible to age (without a saw), just as it was impossible to assess how long it had stood like that, for it was quite dead. But I was captivated by its form, its pose that suggested dance. The trunk was slender and straight for no more than a dozen feet, split into three skinny limbs and six skinnier branches, all of them broken and not a single twig among them. But from where I sat, and as if to disguise or otherwise mitigate its many fatal wounds, three birch saplings cast a net of twigs before it, and as the first of the spring buds were beginning to open, the effect was of a see-through, soft-focus screen behind which the dead tree was frozen in an attitude of dance. I made the poem that begins this chapter for it.

◎ ◎ ◎

The steep path changes character once it emerges from the oakwood. The oak trees had been hemmed in on the lowest slopes for so long by commercial plantation above, the thoughtlessness of the forestry industry. But once the regime changed, once Woodland Trust Scotland replaced Forest Enterprise, once the spruces were felled for good (very much for good) and the mission to re-establish a native Highland woodland ecology was in place, a combination of nature at its most opportunistic and thoughtful hand-planting has been transformative: a mountain birchwood has materialised in less than twenty years. Birds, butterflies, flowers, and a vastly populated new world of invertebrate life have nailed their colours to the mast of the enterprise – their colours, their voices, their eagerness to participate in the born-again mountain. So have a few mammals, notably fox, pine marten, red squirrel, roe deer, stoats and weasels, and I would love to discover that wildcats had found their way back here too, but the fates are stacked against the wildcat in Scotland and the stars are not well aligned, so I am not holding my breath. But all these mammal tribes use the paths to travel, and you find their inches-wide trails through the undergrowth again and again, especially around places they regularly anoint with their droppings. Walk softly, especially in the quietest hours of the quietest days and nights, and sooner or later you will meet them all.

The path responds to this new world variously. Firstly, its verges are freed from the deep shade of the oakwood canopy and eager colonisers have moved in to the sunnier, rainier, snowier, windier world of life beyond the oaks, from Scottish bluebells to eyebright and tormentil, to thistles and inches-high hollies, Scots pine, willow, broom, rowans and alders, among much else. Secondly, if you study the path from time to time (as well as the near and far trees, the skyline, the sky, and that so-critical middle distance – it's a complicated and imprecise trade being a nature writer), you can sometimes read about some of the inmates, especially the mammals. And if you go often enough, occasionally you get very, very lucky.

April was already a week old, and I imagine the weather forecasters were looking for high cliffs to jump from, for it was becoming increasingly hit-or-miss. The temperature in the south of England was fifteen-to-eighteen degrees, and here it was three-to-four. On the question of the icy rain moving through Scotland consolidating into snow, the Met Office seemed to have acquired an approach wondrously articulated by an old hill man I used to know as: "Maybe aye, maybe hooch-aye". Driving across the Highland Edge to Glen Finglas, it became increasingly clear to me that I had unwillingly embarked on an exercise in catching up with a weather front, and by the time I had walked up the path to the point where it cleared the oakwoods I was already deep in that twilight zone between rain and snow and walking steeply uphill…

I stopped to examine a skinny black hieroglyph in the short grass at the edge of the path, and I recognised it as the signature of a pine marten. Then I noticed there was another one on the far side, and just where one of those trails of bent-over undergrowth shimmied away deep into a swarm of birches so dense that the prospects of easing myself through it were not good. This second marten dropping was markedly fresher than the first. How long gone, I wondered. So I listened hard to the wood, and apart from the soft hiss of the hybrid snow-rain and the mutter of a hill burn, there was nothing at all to listen to. Not so much as a syllable of birdsong. This upside-down and back-to-front spring, which had shown up in February with the mistle thrush, had been bludgeoned into submission by the turbulence of March and vanished off the face of the Earth. Not a shred of new green clung to a single tree. In the tussocks beneath the trees, where the trail burrowed into the understorey of the wood, there were puddles of melted snow, and here and there were pockets of unmelted snow where the overhead cover was thickest. Everything about these thousands and thousands of birch saplings and the thickly grown grass-and-heather-and-fern-and-berry-plant hillside they grew out of was utterly cold and utterly sodden.

I walked on but I had taken no more than four or five steps when something stirred a couple of yards from my left boot and just inside the trees, something that had been there all the time and only its stillness had prevented me from seeing it. It was broad and brown and blurred, and my immediate first impression was that it had jumped a vertical yard into the air from a standing start, or even a lying one.

My immediate first impression was quite wrong. It had *flown* a vertical yard into the air from a standing start, or a lying one. Then it snapped into focus and it was a woodcock. First of all, how do they do that? How do they generate the lift for that vertical take-off out of such cloying surroundings? Second of all, how do they learn the exquisitely articulate horizontal flight that followed? I had crouched again to watch it go, and I saw perhaps the first dozen yards of its flight. The gaps between the twiggy extremities of so many close-huddled birches varied between narrow and non-existent. So the flight dodged left and right, and first one wingtip dipped and then the other, and in addition the woodcock had to rise and fall every yard of the way, yet the flight was fluent and unfaltering.

It left a wake, for its wingtips (and I daresay its head and beak and breast, but especially its wingtips) flicked countless twigs as it flew, and every twig unleashed a shower or a cloud of brilliant droplets backlit by a low and wan and appropriately watery sun. For seconds after the woodcock vanished, the trees and their shed cargo of icy water hissed and shivered and shimmered. It was over and done with almost at once. It was and it remains simply one of the most beautiful things I have ever seen.

As the cloud slipped lower and I climbed higher, the rain yielded to steadily falling snow, so I spent more time watching the path, and there were more pine marten scats and more fox scats than I had ever seen there before, and quite a few of them looked fresh. It was reasonable enough, for if the dense undergrowth of the birch saplings was truly saturated – and it was – it made travel for such as a fox and a pine marten pretty uncomfortable, whereas the wide, rough-surfaced path designed for humans (their boots and their mountain bikes) was easy going for both fox and marten, and on such a day as this, mine was the only car in the car park and I was the only human in that part of the wood. All this I registered, and so I took extra care to move quietly and slowly, for who knows what might be round the next corner?

But first I turned aside for the waterfall. The fall is bridged now, or rather the mountain burn that feeds it is, and some distance above, carrying the new path. But for much of its life, there was just the fall. The crag the fall dignifies with its voices (for it has many voices) is still as well-wooded as ever – oak, rowan, birch, a few shapely pines and shapeless hollies, for the foresters and the grazing tribes kept their distance from its gorge – but I imagine all of that once yielded above the crag to the frumpy tenacity of junipers: stumble across these as you descended from the heights, floundering for your bearings in mountain cloud and snow like this, and it would have been as if you had blundered into an encampment of ghostly dwarfs to add to your troubles.

So the fall was ever a turned-aside kind of place, although now there is a signpost proclaiming a "viewpoint" (as if the whole wondrous mountainside is anything less than a natural festival of viewpoints, hundreds of them, if you have seeing eyes, constantly changing, constantly redefining), beyond which an admittedly discreet made-up path curls enticingly out of sight, although it ends fifty yards later at a railed-off stance carved out of the rock. Behold the viewpoint. Behold the waterfall.

The fall is split in two right at the top by a central pillar of rock. The left-hand fall is narrow and contained by a cleft at first but begins to widen as it emerges, then after about twenty feet it hits a ledge and explodes. The right-hand fall is three or four times wider, but it too is briefly constrained by a second cleft before it hits the same ledge, the falls unite and drape a single white curtain eighty feet down the rock face as if magnetised to it. The beauty of the thing is never less than haunting. When the burn is at its most boisterous, the effect is somehow emotional as well as beautiful. I know many waterfalls, and I like to linger beside them, but none moves me the way this one does, and perhaps it is simply because I visit it so often, although I think there may be more to it than I am willing to concede. I'm still working on it.

Gavin Maxwell wrote in *Ring of Bright Water* (Penguin, 1960) that the waterfall near his house was "the soul of Camusfearna". Another of my favourite writers, Ernest Hemingway, was thinking out loud about the soul in *True at First Light* (Arrow, 1999). He was in Africa and had been unable to sleep, and remembered a line of Scott Fitzgerald:

In a real dark night of the soul it is always three o'clock in the morning.

That got my attention because although I have no concept of what a soul might be or whether it exists, I have my share of three-in-the-morning darknesses. Hemingway was casting about for what the concept might be "in the terms that I believed" and came up with this:

Probably a spring of clear fresh water that never diminished in
the drought and never froze in the winter was closest to what
we had instead of the soul they all talked about.

And there was that line of Mike Tomkies about the
black-throated divers being "the true wild spirits of the
magical loch".

It seems to me that, in this context at least, this reaching
for a soul/spirit is a way of literally coming to terms with a
quality of beauty or primitiveness or wildness in nature that
we struggle to put into words. Considering the accumulated
skills of the writers I have just summoned in the matter of
putting things into words, and considering that none of them
is convincing in the use of the words "soul" or "spirit" (and
they all assert or imply that religion has nothing to do with
it), is it not just symptomatic of an acknowledgement that
there are moments in our dealings with nature that we are
unable to articulate because, as a species, we have distanced
ourselves too much from nature, and the animal bonds which
would once have bound us into nature's family have loos-
ened and frayed to the point where we occasionally glimpse
the ultimate fear, which is that there may be no way back? Or
there is this from Barry Lopez in *Arctic Dreams*:

The land is like poetry: it is inexplicably coherent, it is tran-
scendent in its meaning, and it has the power to elevate the
consideration of human life.

I like that thought very much.
I don't live within earshot of the Glen Finglas fall,

although I do sometimes dream about it, and I treasure moments I've spent there. I have also thought a lot about the kind of place it must have been when the land was wholly wild, before the forestry industry, before the overgrazing of the landowning regimes that preceded that, before hunting kings offered bounties on wolves…and perhaps 5,000 years ago when the native woodland was at its height and that country was at its most biologically replete, and that fall still fell. Something of that continuity inhabits the ambience of waterfalls perhaps to a greater extent than any other landscape feature. We are drawn to them in the first place because of their aesthetics, but we linger because they command us to with their inexplicably poetic coherence.

At the Glen Finglas waterfall, a rock wall near the viewpoint accommodates an alder. It grows horizontally from a crack in the rock. How deep within the rock's fissured and fractured innards this perverse tree has propelled its roots is a question to which I would love to know the answer, but unless an act of God or geology were to burst the rock apart, I never will. Its very existence astounds me. It has no trunk, but grows horizontally into space in two thrusts, apparently seamlessly joined at the root, one above the other. The upper one curves slightly upwards as it reaches into the void, the lower one's downward curve is almost a mirror image. A fragile tracery of a handful of slender branches and twigs sprouts from the end of each, and you must look at much of the fall through that tracery. It is illogical to think of it as having the slightest prospect of a long life, yet it was well established when I met it first about fifteen years ago, and it lives on undiminished. Thirty years old? Fifty? Hugh Johnson wrote in his beautiful book, *Trees* (Octopus, 2010), that "…the alder is the streamside tree of neglected places, undemonstrative, a good foil to feathery willow…" and although the thought appears in a caption to a photograph of Oxfordshire, it is no less appropriate to this undemonstrative little outcrop of reincarnating wilderness in Highland Stirlingshire.

Sometimes the survival instincts of nature in the most unpromising of circumstances is a humbling thing to witness. As I watched, the snow thickened and began to lie on the topsides of the two pincer-like limbs. It was April, it was cold, it was snowing, but at the end of the skinniest twigs, little cones awaited, with patience that would shame a saint, the day that would finally dawn that permitted cone to become bud to become leaf. Cones? As if it didn't have enough to put up with, the alder is the only deciduous hardwood in the land to bear cones.

I took my leave of the waterfall and its heartstring-tug of a setting, stepped away uphill among Scots pines. These were, mostly, survivors from the forestry era. When the spruces were clear-felled to make way for nature, these upstart pines, wind-and-bird-borne from higher outposts above the old plantation, danced for joy in unaccustomed sunlight, bowed and bent themselves to the tune of winds they had never previously known. As they had grown from seedlings, they were quickly outpaced by the surrounding spruces in their vertical race for sunlight, so at their own pace they abandoned all notions of limbs, of girth, and grew branchless, straight and tall, hemmed in on every airt other than up. Then one day the clear-fellers moved in and at a stroke (at many, many strokes) freed them from their imprisoned youth. They would never be the handsomest of Scots pines (their only foliage is still just a meagre crown), but they are sources of seeds, and sources of seeds are the lifeblood and the diversity of nature's renaissance.

And there, where the viewpoint footpath rejoined the main path up through the reserve, one more pine marten scat gleamed dully among newly fallen wet spring snow-flakes. That scat had not been there when I arrived. I had an idea and suddenly my day had a purpose. I turned back downhill to where I had seen the little trail that fed back into the trees, the trail whose entrance was marked by two scats on the main path. If I was right about increased marten activity on the main path because of the cloying saturation of the understorey, perhaps a little stillness at a known cross-roads might produce results.

The trail took some finding, mainly because I had not taken enough trouble to pinpoint it on the way up and because close examination revealed a dozen such trails, all of them crossing or ending at the main path. Finally, I found what I thought was the right place and earmarked a pine sapling about twenty yards further down the path which could screen me while I watched. It was the longest of long shots and conditions were not conducive to a long vigil. As it happened, I didn't need one.

I reached the pine, turned to check the view, and I was still in the middle of the path when an adult vixen appeared from the trees on the left of the path and stopped, facing the trees on the opposite side. Seconds later, and from the trees on the right of the path, a pine marten appeared, paused half-hidden on the verge, then stepped forward towards the fox. It is no exaggeration to say that they met nose to nose. There was no sound, no contact; they simply met, scented each other, then moved on. The marten crossed the path and vanished into the trees on the left. The fox looked – for the first time – down the path to where I still stood. Then it looked back over its shoulder in the direction from which it had appeared and into which the pine marten had disappeared, looked back at me, turned uphill and walked away up the path. Ten seconds later there was nothing on the path but two pine marten scats and snowflakes.

Did they know each other? Almost certainly. Neither had a good reason to pick a fight with the other. So they recognised each other by scent, as in a greeting, established that all was well and moved on.

I have seen a lot of foxes. I have seen quite a few pine martens. I have met people who see both regularly. I have never heard of anything quite like what I had just seen. I turned and walked quietly back downhill. At my back, the snow drifted lightly across reincarnating forest. I looked back once to watch it, and as I watched I told myself that there is very little we can do that is more worthwhile in terms of repaying our debt to nature than bringing a half-dead forest back to life.

Swan Song for
a Backward Spring

THE PAGE IS PATCHED with shadow and sun, a small wind moves the shadows around the paper. Distant voices flow nearer, ebb further, none of their words and shouts has meaning except when a child cries.

Be still.

Permit only the writing hand to move.

Feel the late May warmth distil down to the here-and-now writing of the hand, the shadow dance. The distant voices have become a kind of soft focus that helps to sharpen me; they go further back but not quite out of earshot, while my eyes track the words that imprint the page beyond the pen, my own inky spoor. The voices are replaced by the one in my head, the one that articulates my own personal first commandment in the matter of doing this nature writing job:

Write in the very now where you find yourself. There is no substitute even in divine inspiration for the touch of the moment, the touch of daylight on the dream.

They are the words of Margiad Evans, and I never tire

of quoting them to myself and to other people who are interested in nature writing and how it works. That's how.

The in-and-out sun is back. Just before it lit the top of the page I felt it warm my bare foot. Sometimes I think this is the very best time to write, if there is a small wind to take the edge off the sun's true heat. All I ever need then is somewhere to sit, something to write on, something to write with, something to say. An hour passes, leaves no sign it has ever been. Hours do that when I write this way.

A bird's chirrup sounds nearby.

I think "greenfinch" without looking up.

I know birds. Some birds.

Curious: the sounds of nature infiltrate the writing mind long after I have ceased to notice the many far-off human voices and the few that occasionally wander nearby, chattering. These, I noticed before, veer away self-consciously at the sight of the figure on the sun-and-shadow seat, writing. I lean closer to nature, and she to me, so that our heads touch, and I keep the people voices at meaningless distance. Except when a child cries.

A blackbird has begun. I don't remember him beginning, but now he is making something out of layers of song, using long pauses between fragments of his repertoire, making his music make sense. In its way, it is the same thing I do myself, so my writing listens to the blackbird, admires how he does it.

A crow brays four times and is silent.

What was that all about? Was that his idea of music? Did he make it make sense?

And did the still-singing blackbird notice, acknowledge, comprehend? Probably.

And does the blackbird now sing more blackly as a result of the crow? Probably not. The crow does not sing more tunefully as a result of the blackbird.

A child – a boy of about six – leads his own small tribe out of the bushes, two older sisters and a mother follow, too close for my comfort but not his. He sees me and shouts: "Hiya!" from ten feet away. He shouts with a big, open smile. He reminds me at once of my six-year-old self. I was a cheery, hail-fellow-well-met six-year-old too, eyes the uncluttered blue of the North Sea, hair of fine grass bleached fair by the east-coast sun and its compatriot winds, browning skin of freckles. My first friend was the wind. My first enemy was the haar that hove to beyond the cliffs in the evening, swept ashore in the night, silently stole the known world, stole even the voice of the sea. My favourite among first words was "Kittiwake!".

My parents were no one you ever heard of. He moved through the world straight as a ploughing horse, she moved through it singing. When they linked arms and walked, she sang at his tempo. They are long dead now, but when I think of them it is in the springtime when the sun set on them in my eyes and the sun rose over the sea. Oh, have you never seen the sun rise out of the sea? It really is quite fine.

So my own smile warms my shadowed face and says to a smiling six-year-old boy:

"Hiya, yourself!"

I must look harmless enough, for although his sisters ignore me his mother smiles wordlessly but directly at me then herds her small, happy brood ahead of her like a good collie, quietly, no fuss, then out of sight. They looked good together. Her smile lingers where she vanished, like a scent. Likewise my six-year-old self.

They are replaced by a squirrel, a grey one – incomer, infidel, too tame, too cocky, too in-my-face for its own good. It leaps onto a tree and adheres there where it lands, halfway up the trunk. Good trick. Then it spins through 180 degrees from facing vertically upwards to facing vertically downwards. Better trick. But even in that position it contrives to look up, staring at me, begging for food with its eyes. I have advice for such a squirrel.

"The whole garden is food for you. Bugger off."

Off it buggers – still nature though.

The blackbird (I have just noticed him again suddenly, as if at a great distance) sings on and on from pause to pause, still trowelling layers of song onto what it is he makes, still making sense.

My resolve to stop writing for a while, to close my eyes and listen to him to the exclusion of all else, is confounded by the squirrel. It didn't bugger off at all. It canters across the grass towards me. When it stops, it is eyeing me up from a yard away. I resent its scrutiny. It doesn't smile and say: "Hiya." Its stare is not a greeting but an unspoken demand: "Gimme."

It stands two-footed for endearing effect. It won't work.

"Bugger off."

It steps nearer, eyeing up my feet. It seems not to know that these kick. At my first foot-flicking gesture it jumps back a yard, spins in its own length, four-footed spin, stands, feels in a waistcoat pocket with one hand, then buggers off, this time as requested.

My eyes return to the hand, the pen, the page, the shadows, the sun; my mind to the writing while the sunlight still illuminates the dream.

Suddenly, the swan.

Wingbeats up there in the eye of the sun, so heard rather than seen. That sound of wonder, that exquisite throb, four in the bar like a Freddie Green guitar, crescendo then diminuendo then gone. Where do you fly, swan? Why do you always gatecrash my quietest hours with such grace, such certainty, such timing?

Who's Freddie Green?

Played guitar in the Count Basie Orchestra for forty years, the truest pulse in jazz. I heard him on the radio last night, just before the news. I only listened to the news because there was more jazz after the bulletin.

In the bulletin, a girl aged eleven, in Alsace, France, hit by a car then thrown into a lake by its driver while she was not yet dead. She died of drowning and she died of her injuries. Eleven years old and she died twice. So the voices reach me again only when a child cries, or a six-year-old boy shouts: "Hiya!" and grins.

The bulletin finished with the sport and the weather, then the jazz kicked back in, Bessie Smith, "St. Louis Blues", 1920-something. The voice is crying out of its time like a child who died twice. Oh, but the cornet fill-ins make me smile; they shine like a sun rising up out of the sea.

What did they make of you, Louis, when you first stepped out of the shadow of King Oliver and began to play like that?

Sometimes I think it would have been good to have been

in at the beginning of jazz, to feel all the tentative stirrings coalesce at last in a Louis Armstrong cornet solo, to see and hear all jazz gather behind its one true Messiah, confident for the first time of its direction. Sometimes, though, it's enough to let it lift me from the depths of the news bulletin, so that I think:

Yes, we're capable of that with the eleven-year-old girl, but also capable of a Louis Armstrong solo, of a Freddie Green that plays rhythm guitar with the pure certainty and the natural grace of a swan's wings. I wonder if Louis ever played with Freddie? I would like to have heard that gig.

Louis, in my ears, you make your music make sense. You and that blackbird.

The radio last night had followed Louis and Bessie with Billie Holiday, "Lover, Come Back To Me", Lester Young's tenor her foil, as Louis's cornet was to Bessie. It was too much then, tears through the music for an eleven-year-old girl who died twice in a lake in Alsace, France. After I pressed the off-switch, a blackbird beyond the window had brimmed the silence. We all try – blackbird, Bessie, Louis, Billie, Lester, Freddie, swan, even me on my good days – we all try to make our music make sense. And a six-year-old boy bursts from a thicket of bushes towing his preoccupied sisters, his attentive mother.

"Hiya!"

"Hiya, yourself!"

Then the wordless music of her smile, the swan song for a backward spring.

Acknowledgements

THE WRITING OF SETON GORDON has been a close companion for many years now. A clutch of his books are among the most thumbed in my bookshelves. In the course of writing this book, I have been particularly grateful to *The Land of the Hills and the Glens*, *The Cairngorm Hills of Scotland* and *Afoot in Wild Places*. And every time I write about golden eagles, I am indebted to the groundbreaking work he had put in place long before I was born. Every Scottish nature writer over the last hundred years will surely acknowledge him as the foundation stone of their art, and who knows how many thousands of inquiring minds and watching eyes he has opened through his books. I doff my cap in profound gratitude one more time.

I am doubly grateful, therefore, to James Macdonald Lockhart, not only because as Seton Gordon's great-grandson he granted me permission to quote freely from those masterpieces of 20th-century nature writing, but also because he and his publisher, Fourth Estate, gave me permission to quote from his own fine 2016 book, *Raptor*. James and I met at a book event in St Andrews a couple of years ago and, as you can imagine, we had a great deal to talk about.

One of those benevolent coincidences which sometimes attend my writing life came in the form of a wolf-related email while I was writing the wolf pages of this book. It was from Raymond Besant, an Orkney-based wildlife film-maker, and I am grateful to him for allowing me to quote it here.

Those master craftsmen of the writing art, Gavin Maxwell, Aldo Leopold and Barry Lopez, appear regularly in the acknowledgements pages of my books and they are here again; my gratitude to them for their work and their inspiration is eternal, and I return to their books constantly.

Thanks to Birlinn for permission to quote from Robert Hay's impeccable history of Lismore.

Thanks, too, to old friends Sheila Mackay (for allowing me to quote from her admirable little book *Lindisfarne Landscapes*); the late Mike Tomkies, who taught me so much about eagles; and the late Syd Scroggie, who was a shining example to all of my generation who grew up in Dundee with our eyes turned northwards towards the hills of Angus.

I continue to struggle laboriously with my poetry, and although a few examples have crept into this manuscript I am forced to acknowledge my limitations in the company of the masters, notably Norman MacCaig, whose "Landscape and I" is a wonder to me.

Finally, to Sara Hunt of Saraband, her editorial director Craig Hillsley, and my literary agent Jenny Brown: thanks to all of you for hanging in there and for looking after the work so well; and thanks to Jei Degenhardt, for proofreading.

JIM CRUMLEY HAS WRITTEN more than thirty books, mostly on the wildlife and wild landscape of his native Scotland. His work has been shortlisted for prestigious awards such as the Wainwright Prize and the Saltire Society Literary Awards. Jim is a widely published journalist and has a monthly column in *The Scots Magazine*, as well as being a poet and occasional broadcaster on both radio and television.

Also in Jim Crumley's seasons tetralogy

Longlisted for the Wainwright Book Prize 2017
A pilgrimage through the shapes and shades of autumn

IN AUTUMN NATURE STAGES some of its most enchantingly beautiful displays; yet it's also a period for reflection – melancholy, even – as the days shorten and winter's chill approaches.

Charting the colourful progression from September through October and November, Jim Crumley tells the story of how unfolding autumn affects the wildlife and landscapes of his beloved countryside. Along the way, Jim experiences the deer rut, finds phenomenal redwood trees in the most unexpected of places, and contemplates climate change, the death of his father, and his own love of nature.

He paints an intimate and deeply personal portrait of a moody and majestic season.

> "Winter is the anvil on which nature hammers out next spring.
> Its furnace is cold fire. It fashions motes of life.
> These endure. Even in the utmost extremes of
> landscape and weather, they endure."

DURING WINTER, DARK DAYS of wild storms can give way to the perfect, glistening stillness of frost-encrusted winter landscapes – it is the stuff of wonder and beauty, of nature at its utmost. Here, Jim Crumley ventures out to experience firsthand the primitiveness and serenity of nature's rest period.

He bears witness to the lives of remarkable animals – golden eagles, red deer, even whales – as they battle intemperate weather and the turbulence of climate change. And in the snow Jim discovers ancient footsteps that lead him to reflect on his personal nature-writing life – a journey that takes in mountain legends, departed friends and an enduring fascination and deep love for nature. He evokes winter in all its drama, in all its pathos, in all its glory.

Also by Jim Crumley

NATURE WRITING
The Nature of Winter
The Nature of Autumn
Nature's Architect
The Eagle's Way
The Great Wood
The Last Wolf
The Winter Whale
Brother Nature
Something Out There
A High and Lonely Place
The Company of Swans
Gulfs of Blue Air
The Heart of the Cairngorms
The Heart of Mull
The Heart of Skye
Among Mountains
Among Islands
Badgers on the Highland Edge
Waters of the Wild Swan
The Pentland Hills
Shetland – Land of the Ocean
Glencoe – Monarch of Glens
West Highland Landscape
St Kilda

ENCOUNTERS IN THE WILD SERIES:
Fox, Barn Owl, Swan, Hare,
Badger, Skylark, Kingfisher, Otter

MEMOIR
The Road and the Miles

FICTION
The Mountain of Light
The Goalie